Systems in Action

A MANAGERIAL AND SOCIAL APPROACH

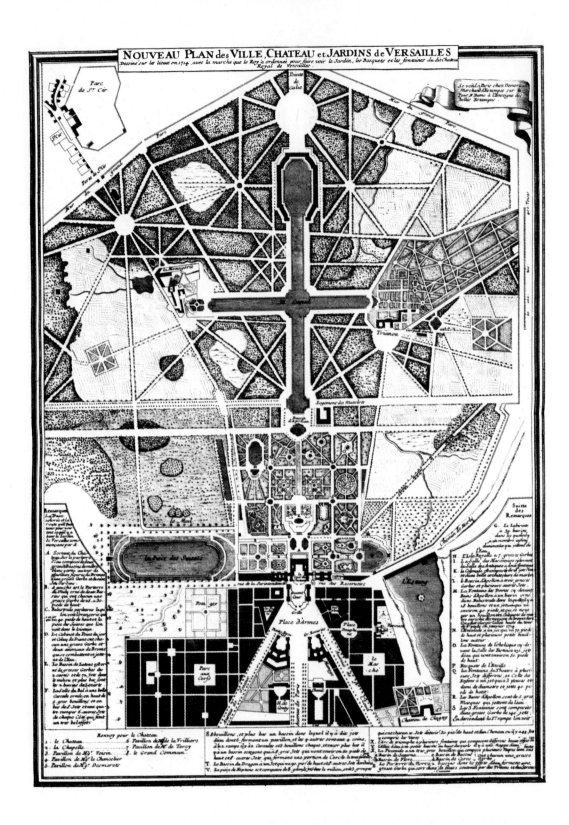

Systems in Action

A MANAGERIAL AND SOCIAL APPROACH

Joseph Allen

Bennet P. Lientz
University of California, Los Angeles

Goodyear Publishing Company, Inc.

Santa Monica, California

Library of Congress Cataloging in Publication Data

Allen, Joseph, 1944–
    Systems in action.

    Includes index.
    1. System analysis. I. Lientz, Bennet P., joint author. II. Title.
T57.6.A44   003   77-16546
ISBN 0-87620-856-1
ISBN 0-87620-858-8 pbk.

Y-8588-9 (p)

Y-8561-6 (c)

Current Printing (last digit)

10 9 8 7 6 5 4 3 2 1

Printed in the United States of America

Production Editor: Linda G. Schreiber
Copy Editor: Margot Tommervik
Cover and Text Design: Design Office / Bruce Kortebein

———————

To Mary and Martha, who let us choose the better part.

———————

Contents

List of Figures

Preface

This book is addressed to people who solve problems. It's also addressed to people who should be solving problems, but aren't. It's addressed to the student of management, of social science, of engineering, of public administration—to anyone who needs a logical framework to help deploy the modern analytical tools that we've seen proliferate since the 1950s.

The aim of this text is practical; it's down-to-earth; it's everyday. We want to teach a way of approaching problems with a keen eye toward solving them efficiently and successfully. This is not a book of theory, nor a guide to computerized systems. It's an approach to a way of thinking that has, we believe, wide-ranging validity in today's world. In many important ways, the systems approach is simply a commonsense approach to problem solving. Its steps are logical, phased, and geared to avoiding failure in any of the applications possible.

We have attempted, throughout the text, to concentrate on the human aspects of the systems approach rather than on the technical. We have included sections on topics previously omitted from discussions of this approach. Thus, we've included sections on interviewing techniques, on traps and pitfalls that point the way to failure. Although we've used examples from many disciplines, we've chosen a single case study to function as an organizing factor throughout the text.

That single case study is an inventive (rather than factual) view of the building of the Palace of Versailles. Our protagonists are Louis XIV of France and his court, especially his great minister, Jean Baptiste Colbert. Our aim here is not to present an accurate historical view of these men and their times, but to illustrate *what could have happened* if Louis and Colbert had used the modern systems approach in their great undertaking.

We chose the Versailles example for several reasons. As the reader will soon become aware, it is a multifaceted example, employing political, economic, and physical solutions to a multitude of problems. We believe our discussion of the history of France to be basically correct and defensible, but we must emphasize that we have bent the situations to our own ends and to the ends of the modern student of systems, rather than to the goal of historical accuracy.

This book is organized around the system life cycle, which we view in seven stages:

1. Initiation, when an existing system fails or is close to failure, and the need for a new system first becomes apparent to the people who will use it.
2. Feasibility, in which a preliminary survey of the problem, the environment, and the possible alternatives is made.
3. Analysis, the first pass at a detailed analysis of the problem that requires a systematic solution.
4. Design, in which the solution is drawn up, detailed, and readied for construction.
5. Building, which speaks for itself.
6. Installation, which includes a discussion of the problems of "switching over," as well as the essential process of training and testing.
7. Operation, which we've broken into two chapters—one on maintenance, the other on enhancement.

We've imbued this text with the consideration of trade-offs, of costs incurred and costs avoided, and of potential sources of failure. We've omitted, in many cases, discussion of specific techniques of analysis, which vary greatly from discipline to discipline. We've included important techniques where we feel they are appropriate (including sections on cost-benefit analysis, PERT, and the Critical Path Method).

We've included exercises for discussion after each of the first ten chapters. In these exercises, we've tried to follow a linear progression set up in the chapters themselves. The exercises are in some cases related to specific disciplines—management, sociology, engineering—and, as such, may be of only partial interest to some readers. We've tried to cover all bases, however, and there's something under each important heading for (we hope) every reader.

The reader who finishes this text should have a framework for reference in a host of different problem situations. He should be able to slot in concepts, techniques, and applications from virtually any discipline. The systems approach has been traditionally a proprietary tool of the hard and applied sciences. We've presented it in such a way as to preserve a pathway for the practitioner of the so-called soft sciences as well.

This book is aimed at helping the reader prevent failure in any problem situation, from whatever discipline or industry. We have great faith in the application of the systems approach to problems in a wide range of real-life situations. When the systems approach is viewed—and used—as a way of thinking, rather than as a formalized, carved-in-marble set of procedures, its value doubles, triples, multiplies almost unimaginably. It is our hope that this book will help spread that good news.

We wish to thank several people whose helpful comments and positive criticisms have made it possible for us to write this book. We wish to thank several members of the Graduate School of Management at the University of California at Los Angeles, Dr. Paul Kircher, Dr. Jack McDonough, and Dr. Burt Swanson, for their invaluable help in this regard. Special thanks are also due to Ms. C. U. Greaser, the Rand Corporation, for her comments and suggestions. We appreciate the helpful comments of Ms. A. T. Collins, Mr. Steve Lock, Mr. Chris Jennison, and Mr. C. Gardner Sullivan. Particularly helpful were the valuable comments and encouragement of Dr. David Rodes, Department of English, University of California at Los Angeles. Last (but not least), we must thank the students who have participated in the formation of our presentation—without them, we might still have a book that spoke most fervently only to ourselves.

Joseph Allen

Bennet P. Lientz

Systems in Action

A MANAGERIAL AND SOCIAL APPROACH

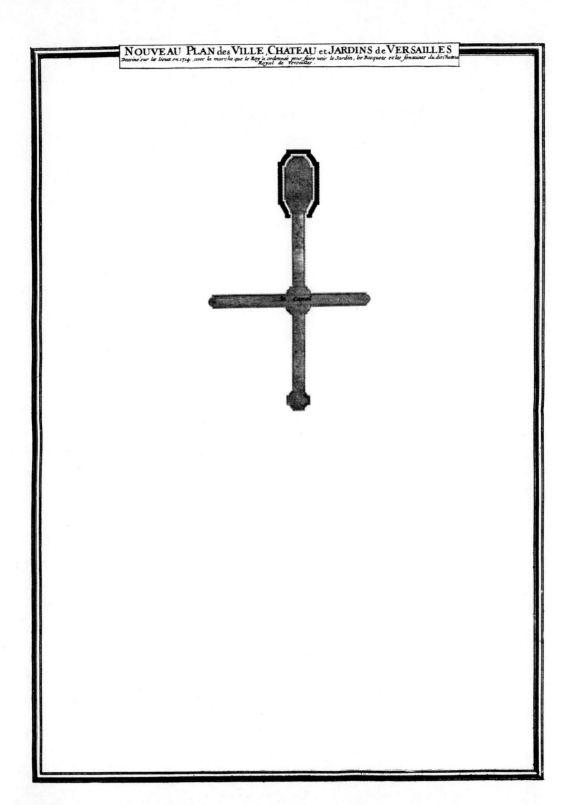

1

Introduction: Birth, Life, and Death

To many people, the term *systems* implies computers. After all, systems are those things that computer folk developed to help structure computer projects.

The fact is that the systems approach is not limited to computers. The field has grown so in the past few years that systems that do not employ computers outnumber those that do. The reason is that the systems approach is not and has never been strictly a hardware (or computer) technique. It's simply a way of looking at—and solving—problems, of whatever ilk and orientation.

What are some of the general characteristics of the systems approach? In a few words, the systems approach is a thorough, phased approach to problem solving. It's an approach geared to avoiding failure (when it is properly used) and to putting available resources to their "highest use." The systems approach never condones overcommitment of resources, and it requires deliberate approvals at several steps during its life cycle.

The systems approach provides a skeleton, a framework within which to apply techniques in solving problems. It evolved as a step-by-step approach because of the nature of computers (which require a logical progression of ideas to produce a logical conclusion), but that step-by-step approach is applicable to any number of noncomputer situations.

In this text, we will work with the systems approach from a noncomputer viewpoint. The framework we'll display is one that can adapt itself to virtually all disciplines. We have chosen case studies to illustrate a number of them.

4   *Systems in Action.*

But it's most important for the reader to keep in mind that we are teaching *systems* and their applications—not any specific field or discipline. Each of the examples and case studies has a specific point to make in the field of systems.

The System Life Cycle

The most obvious, and perhaps most important characteristic of systems is that they occur in cycles. While that may sound complex, it actually isn't. Let's consider the concept of life cycles for a moment, and then come back to the idea of a system life cycle.

A child comes forth from its mother, matures into an adult, dies as an elder. This brief summary of life on a human level is the most basic example of a *life cycle:* a repeating process which provides our planet with human inhabitants. While we aren't concerned with biological life cycles in this text, let's use them for a moment as a basis of comparison.

Taking the three stages of birth, life, and death, apply this basic human form of life to the following: plant life, the duration of a love affair, the practical longevity of an automobile. All these have an inception, a period of life or utility, and an end. But who looks at a shiny new car and thinks seriously about an automobile as a *transportation system,* with a beginning, a middle, and an end (birth, life, and death)? Answer: virtually every potential buyer, who must decide initially about financing, then explore the eventual resale value of the automobile as a used car. Such a car buyer is actually considering the car as a system with a discernible life cycle: a birth (purchase), a life (period of use), and a death (sale or trade-in).

Recognition of this life cycle is basic to the planning or creation of any system in an earthly environment. Nothing starts without a beginning, and nothing exists without an end. Noble buildings crumble eventually (or are abandoned), despite the best intentions of preservers and maintainers. The most fervent love affairs burn down to embers unless their life cycles are preempted by others.

We intend in this text to focus on the life cycles of systems: business systems, administrative systems, social systems, and information systems. The cycles of all of these are basically similar to the cycles of nature and human life: each begins, exists for a time, and dies.

Within these all-encompassing life patterns, there are stages of creation, existence, and death, and every successful system generates at least the hope of a succeeding generation of life, just as in the biological world.

A full and complete understanding of these stages of life will become, through application of the systems approach, a means of ensuring maximum success in creating systems. The principles will remain the same whether the systems be in management, in engineering, in education, in social science, or in any other vocation or discipline. But before we venture into the specifics of life cycle stages, a word about *failure*.

Failure

You almost want to stop reading when you see a heading like this one, don't you? The word carries a taboo that repulses the bravest among us. But whether we like it or not, there it is: failure is a possible result of any system we might attempt to create. A child can be stillborn; a machine can fail to function; a computer program can become obsolete before it is finished; a seed can fail to germinate. An otherwise perfect system can have failed to take into account an all-important environment without which it is useless.

One of the prime reasons for studying the systems approach is to minimize the possibility of failure as a finishing mode. The first step in the prevention of failure is, however, to learn to distinguish it from *death*. Schweitzer, Nehru, and Hammarskjöld are all dead. But their lives were in no way failures. While that may seem an extreme example of death as distinguished from failure, it's designed to make a point. Death is an essential and unavoidable segment of any life cycle.

The marvel of German World War II engineering, the V-2 rocket, is outmoded today; it's a museum piece, a clumsy attempt at something that today has culminated in shots at Mars and beyond. The V-2 had an easily delineated life cycle; at a given point in time (when something better was developed), it was packed away. It died. But, since it was the pioneer in today's space probes, very few people would say that the V-2 was a failure.

Similarly, the earliest computers are past their utility in today's business and scientific community, but they weren't failures: their children and their children's children run our airports, compute our taxes, analyze our blood samples, and calculate the amount of pollution in our air. Death isn't failure; it's simply an inevitable part of the natural life cycle.

Failure is a prevention of utility. It's usually unforeseen; it's virtually always undesirable; it's frequently preventable. And it will happen from time to time.

The easiest way to induce failure is to ignore the basic life cycle and to push a useless system into a situation it can't cope with. Like the old proverb about whipping dead horses, an outmoded system won't pull a new cart no matter what kind of prod is used. Disregarding the inevitability of the life cycle is the most common way to induce failure.

The business world is a great offender in this: it is the most likely to overload a relatively successful system and kill it. A well-made automobile sells well, and the company that makes it gears production up to pump out more of the same. But the automobile was well-made because it was carefully made—and the large output of the overloaded factories no longer consists of the fine machines that sold well. Sales fall; the system fails. The manufacturer failed to take into account that the system he was using was outmoded by the increased demand. When he failed to see that either (*a*) he would have to take orders rather than mass-produce or (*b*) he would have to revamp manufacturing procedures substantially to ensure quality, he ignored the fact that the system wasn't able to cope (unaltered) with the new sales environment.

System overload is a failure method not uncommon in other areas. The suspension bridge that collapses under the weight of too many vehicles is a perfect example of a dead-horse system. Others include the expectation that a new freeway will reduce traffic congestion, the political notion that more people will ever burn less gasoline of their own accord, and the attempt to pull college-level knowledge from a children's encyclopedia.

Failure may be difficult to recognize. It can be disguised as inadequacy or as a series of preventable breakdowns. Worst of all, it can sometimes appear as success. The drug and pharmaceutical industry can supply plentiful examples of failures that were originally hailed as successes (thalidomide comes to mind immediately).

The prevention—and detection—of failure is an exceptionally important part of system design. It is easiest to detect and prevent when systems are designed carefully, with full understanding of the necessary components of the system life cycle and of the environments in which systems are built and installed.

Success

There, that's better. We've dismissed failure. Now we turn to success—which may be a bit more difficult to describe but is much more pleasant to experience. A successful system can be the one that simply poses a whole new set of problems, or it can be the one that does no more than outmode an old system. One Venus probe, for instance, discovered that the planet's surface temperature is much too hot to support life and that its atmosphere is made up of hydrocarbon gases. That information invalidated most previous theories about the origin and potential of the planet. (It also made a very happy man of Dr. Immanuel Velikovsky. Among other predictions, he had forecast the Venus findings twenty years earlier and had been treated as a quack.)

Success, like *failure,* is a relative word. It implies judgment on the part of the speaker. It also implies continual reassessment. There's an old saying about winning battles while losing wars that embodies this concept. After all, the internal combustion engine was considered an unqualified success not many years ago. Now, some people are beginning to wonder.

For all the difficulties in achieving and identifying success, it's the end for which all living systems strive: success in what we do, the accomplishment of our aims. And success must be the object of systems, or there's no point to them. Any system that does not accomplish its goals does not succeed; it fails. And any system which does not fail, succeeds.

Sometimes it isn't so cut-and-dried, though. Success can come mitigated by bits and pieces of failure. A system that accomplishes its major objectives but misses (or partially misses) some secondary objectives is a good example. One thinks almost automatically about federal tax legislation intended to close loopholes in income taxes. Loopholes are frequently closed by such legislation, but unforeseen side effects occur.

So, having faced the possibility of failure and glimpsed the hope of success, we can turn back to our discussion of the life cycle.

The Stages of Life

We said earlier that life cycles are made up of three major time periods: birth, life, and death. That three-part division is useful in discussing systems at large, but it isn't detailed enough to help us build systems. So we've divided up the life cycle a bit more—into seven major stages. These seven stages follow a system from inception to creation to installation and through operation to death. Let's take them one by one.

Initiation

Every system has its beginning in a problem, a problem that is chafing society or one of its members. Perhaps that problem is financial; perhaps it's governmental, commercial, or social. But something is bothering somebody somewhere.

That chafing generates the seed of a system: the search for a solution. After all, the first step in solving a problem is recognizing that a problem exists. Systems are responses to problems, and the problem itself must generate the initial search for a solution.

Folk wisdom has it that "necessity is the mother of invention"; for our purposes we can adapt that proverb to read "problems are the mothers of systems." The electric light was Edison's response to a given set of problems: what to do with electricity to make it a useful tool of society; how to supply clean, smokeless light to a dark world. Similarly, Jonas Salk's development of an effective vaccine against poliomyelitis was the response of a scientist to a problem: how to overcome the alarming rise and devastating effects of a major disease. Both systems worked: Edison found the principles of incandescent lighting; Salk isolated a microorganism that allowed him virtually to wipe out polio in a single generation.

Edison and Salk responded to specific problems by identifying the best probable course of action and pursuing it. Edison's pursuit of a solution took him through a series of experiments and partial failures in a laboratory; Salk's system life cycle involved both laboratory and "real world" testing. Each man created a system to respond to a need; both systems succeeded.

Practically speaking, every problem situation has two components:

1. The problem ("How can I use electricity to produce light?" "How can we eliminate polio?").
2. The solution.

For our purposes, we'll assign roles to each, and presume that one person or function (the "user") detects the problem and seeks another person or function (the "system developer" or "problem solver") to produce a solution. In the cases of Edison and Salk, these functions might be broken down as:

CASE	USER	SYSTEM DEVELOPER/PROBLEM SOLVER
Electric light	Society	Edison (and later his research associates under his direction at Menlo Park)
Polio vaccine	Victims and potential victims of polio	Salk

The system developer functions as a tool of the user, who reaps the benefits and rewards of the successful system (and stands the costs). During the stage of the life cycle we're calling initiation, the user approaches the system developer with a problem, and the system developer surveys the problem, broadly, looking for the outline of a solution ("system"). The synthesis of these two functions produces the beginning of a system. Life has begun.

System Development

Most of us (all of us?) are familiar with problems, and can readily identify the user as a function or role. The concept of system development may not be as familiar. Let's have a look, while we allow the system we've just initiated to develop a bit.

The system development function has one purpose: to respond to a given problem or set of problems. Perhaps the most familiar examples of system development groups are those vocations society refers to as the "Professions," although virtually the whole vocational world consists of problem solvers of one ilk or another. The professions—medicine, law, teaching, and the clergy—are all systems designed to respond to particular types of problems. The practitioners of these professions are constantly developing new systems to deal with new problems.

The medical doctor is a specialist in human biological problems. The lawyer is a problem solver in torts, suits, trials, and the like. The teacher is a practitioner of society's need for indoctrination. The clergyman responds to our problematical longings for ethics and immortality. Each of these people embodies a self-contained system development role. We may approach any of them with a problem and hope for a solution.

In the complex worlds of business, government, the arts, and academia (and oh-so-many others!), system development isn't as easily isolated as in the professions. But it's there, in various disguises. Examples might include the marketing function in a sales organization, a troubleshooting mission in the Department of State, the performing arts councils at cultural centers, and academic senates in universities. When we remove their masks, we find ourselves surrounded by system developers in myriad specialties.

These complex system development groups contain within them various roles. Although the roles frequently overlap in practice, we can subdivide them into three main categories:

1. Management, the decision-making role, which allocates resources, approves progress, and appoints personnel.

2. Project leadership, which oversees the actual development of problem solutions (systems).

3. Analysis and development, which, under the guidance and budgetary constraints set out, actually does the work of developing the system.

Figure 1.1 illustrates the relationship of these three roles within the system development group. Note that we have chosen a pyramid shape for this figure; it reflects the proportions of such groups (in terms of number of people and number of duties) as well as their reporting relationships.

FIGURE 1.1

Roles in System Development

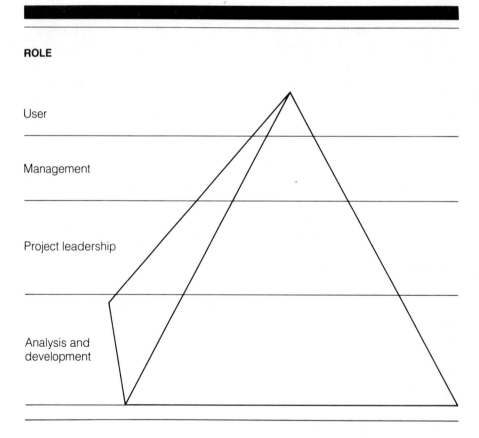

ROLE

User

Management

Project leadership

Analysis and
development

In a business environment, the *management* and *project leadership* functions are performed by members of management, while the *analysis and development* role is filled by professional and technical workers and the labor force. We can break down the examples listed previously into these three roles, as we have in figure 1.2. Similar breakdowns are possible in the development of any system, although in less complex systems more than one of these roles may be fulfilled by one person.

FIGURE 1.2

Examples of Roles

SYSTEM	MANAGEMENT	PROJECT LEADERSHIP	ANALYSIS AND DEVELOPMENT
State Department troubleshooting mission	President	Secretary of state	Special ambassadors
Marketing department	Marketing manager/vice-president	Specialist in specific product marketing	Market analysts, pollsters, EDP, advertising department
Performing arts council	Board of directors	Program director, fund-raising chairman	Press agents, accountants, set designers, and so on
Academic senate	Senate executive committee	Committee chairman	Task forces

With the four roles (user, management, project leadership, and analysis and development) embroiled in the development of a system, we can return to the incipient solution we mentioned in our discussion of the initiation stage and allow it to confront us as a question of feasibility.

Feasibility

In the human life cycle, there's a nine-month feasibility period, during which the child develops inside the womb. At the end of that period, the child is a viable human being, and is born. A system operates much the same way.

You will recall that a few pages ago we left a user with a problem chafing him—generally causing problems. We said that this chafing led him to seek a solution from a system development organization. There was an initial meeting of minds, and the system was initiated. Life had begun in some form. But, just as nature and the human body take nine months to decide finally on the viability of a human being, the system development group spends some time deciding on the feasibility of a specific project.

Once Thomas Edison or Jonas Salk recognized a specific problem and decided on a tentative response (Edison: "Do something with electricity"; Salk: "Develop a vaccine"), a period of exploration began, which we will term the feasibility stage of the life cycle.

In Edison's case, the feasibility stage took the form of probing into the manner in which filaments glow when a current passes through them. In Salk's system, the feasibility stage involved determining that a specific microorganism causes polio, a microorganism that might be susceptible to a human antibody reaction. Both men found avenues to follow: Edison found that wire filaments glow with enough magnitude to generate measurable light when subjected to a current of electricity. Salk isolated the virus he was looking for. Salk isolated the organism which produces infantile paralysis in human beings; having thus identified his "enemy," he was able to proceed.

In the systems we are likely to encounter, the feasibility stage is that part of the life cycle during which the problem is first studied in depth, with an eye to eliminating all alternatives but the one most likely to succeed. Facing the problem of a budget deficit, the performing arts council would pass through a feasibility stage of studying all the possible methods of raising money: soliciting governmental grants, preparing public appeals and foundation proposals, making cuts in expenditures or reductions in programming— whatever the alternatives facing them might be. From their study of these alternative possibilities, the analysis and development (A&D) group at the performing arts council will respond with recommendations for specific courses of action. A system is born, a system that will attempt to survive, mature, and function.

Analysis

Once a path of action is recommended for a specific problem, the A&D group undertakes a study of the specific alternative selected for action during the feasibility stage. This period of study is the third stage of the life cycle: the analysis stage.

Just as the newborn infant must explore the problem (life) set for it, the analysis stage involves investigating thoroughly all aspects of the course of action dictated during feasibility. The State Department mission sets up a fact-finding team to report on the total situation in the troubled territory. That fact-finding team reports on history, socioeconomic conditions, political status, military strength, possible involvement of foreign powers, important personalities, natural resources, and balance of trade payments. The degree to which the analysis stage is successful in investigating all the possibilities dictates the degree to which the system will succeed.

The work set out for the system development group at this stage is pivotal. It is the basis for all parts of the life cycle to come. It consists of fact gathering (woolgathering?), fact digestion, and output in the form of a body of information from which a design can be made. It leads naturally into the fourth stage of the life cycle: design.

Design

During the fourth stage of development (design), the organism (whether human or figurative) must put together a design for its eventual purpose. It must base that design on the knowledge it has gathered during its analysis of data, although the input of information continues to build throughout the life cycle.

During the design stage, Thomas Edison had to map out a practical method for making wires glow in the dark—a method that didn't require a whole laboratory to make it work. Salk had to decide which of the many methods of making vaccines would produce the most effective vaccine for infantile paralysis. Both men designed systems to accomplish their aims.

Details of the structure of the system are set out during the design stage; these details will enable the construction workers in the next stage to build the edifice the way it must be built. In an electronic data processing environment, design dictates the ends and parameters of specific programs. In the case of an academic senate trying to formulate a policy on professorial tenure, the subjects to be discussed and the manner in which they are to be handled must be decided.

Edison decided on a remote electrical generation system, a pattern of conducting wire, and a filament encased in a glass globe. The building stage followed closely on the heels of that decision.

Building

Perhaps the most important single aspect of the building stage of a system (aside from the actual work of construction) is the *commitment to a specific course of action.* In a business environment, during the building stage, the marketing department must embark on an outlay of funds for a marketing plan. It must commission newspaper ads, television commercials, direct-mail campaigns,

Le Corbusier's modern Secretariat contrasts sharply with traditional Punjabi life during the building stage of the new city of Chandigarh in India.

and point-of-sale aids to accomplish its aims. The academic senate must write its policy statement. The State Department must begin constructing its public course of action. The performing arts council must announce a season and put it together.

The commitment to a plan is virtually total by the end of the building stage, and the largest single outlay of funds, energy, and resources is involved in bringing it to fruition. A human being in this stage of his life undergoes intensive schooling or training, intensive exposure to the world outside the family, and intensive preparation for independent living. At the end of this period he has very little chance of returning to the family environment for regrouping or replanning. Any alterations in his life plan must be made in the middle of the rising edifice he has decided to create. A good example of the commitment is the college senior who decides to change his major from sociology to French literature: with few exceptions his parents won't take the additional three years of undergraduate work kindly, and he'll probably have to choose between paying his own way and living with a sociology degree for the time being.

During the building stage, Edison built a light bulb. Salk developed a vaccine and tested it on animals.

Installation

The day after you graduate from college is a scary day. It's the day that the want ads in the paper become a threat instead of a throwaway. You're faced with making the machine you've been building all your life run. This is what might be called life's installation stage. Although our analogy may seem a bit weak, it's clear that now you must sink or swim as you are. Watch out!

During the installation stage, a system is turned back over to the user—the fellow who had the itch in the beginning. It's up to the organization that developed the system to make certain that all parts are operational and that training manuals are available where needed.

It was at this point that Edison gave the light bulb to the world and that Salk announced the cure for polio. There's no shortage of examples of failure at this stage—which brings us to another short discussion of failure. Remember Ford's grand design for an automobile revolution—the Edsel? Remember Napoleon's potentially greatest victory—Waterloo? It's foolish not to test for failure prior to installation. Happening successfully at this stage of system development, however, were such triumphs as Lindbergh's setting down at Le Bourget airport near Paris, the successful incursion at Normandy beachhead during World War II, and the adoption of the United States Constitution.

During this phase, the performing arts council offers its season to the public; the academic senate offers its policy statement to its members for approval; the State Department offers its plan to the President; and the marketing department buys some air time on CBS.

Ready or not, here we go!

Operation: Maintenance and Enhancement

With the system proven successful, it's now in operation, back in the hands of the user. The new ballet is being played to sellout houses at the cultural center. The tenure policy is protecting academic freedom. The State Department is dealing effectively with country X. The product that the marketing department is pushing is selling like hotcakes.

Operation is the last phase of the system life cycle, and it can go on indefinitely in some systems (like Edison's or Salk's). In one-shot systems (such as a moon landing), it either works or it doesn't. During the operation stage, *maintenance* is carried out as needed; *enhancements* are made where they are essential or desirable.

What have we said?

We've delineated seven stages in the development and operation of the system life cycle: initiation, feasibility, analysis, design, building, installation, and operation. We have shown how each one in the sequence is entirely dependent on the success of the stage before it for its own success.

We have discerned four roles at work in the development, operation, and maintenance of systems: user, management, project leadership, and analysis and development. We have discussed failure and success as alternative modes of operation and have preferred the latter to the former.

We have followed Edison and Salk in their pursuits of two system solutions that have changed the face of our world. More important, though, we have delineated a *system life cycle* which is applicable to systems of all kinds: birth, life, and death.

What more do we want?

We want to discuss briefly some applications of the systems approach. The systems approach has traditionally been the somewhat arcane territory of the computer-oriented sciences; it has spread to management and social science. It has had some applications in engineering and some in mathematics and economics.

The systems approach is a useful and practical way of solving problems, any problems. It isn't mystifying or occult, and it doesn't require FORTRAN or COBOL. It is a very simple, logical way of approaching problems; and it's also very precise in quantitative situations.

The soft sciences have for years suspected that they too could use the tools of the hard sciences. Whether or not the soft sciences will ever make that transition remains to be seen, but the systems approach is a bridge. It's a down-to-earth, nuts-and-bolts way of ordering things that don't just order themselves.

Where are we going?

We're going to build something. We want to test the validity of what we've said, and build something the world will remember. Since we're going to have to live with what we build, we've decided to make it comfortable. We've decided to build a palace. We've chosen a sleepy village outside seventeenth-century Paris, and we've moved back in history 300 years.

There's a river that runs through this village, called the Galie. Actually, it's little more than an enlarged creek, but it's the largest creek around, and it's given its name to the whole territory around here: the Galie Valley.

The village we've chosen has been around for a long time, and has been a center of the beef trade between Normandy and Paris. The lordship of our village has changed several times during the last one hundred and fifty years, but it was finally acquired by the father of the current king as the site for a hunting lodge.

The date is May 4, 1643, and Louis XIII of France has died. The new king is a child of five, Louis XIV, who yearns to be like his father in all things. He particularly likes his father's hunting lodge in our little village. The lodge is a modest affair with a small garden. But it ought to be somewhat enlarged in a setting so lovely.

The name of the village is Versailles.

Louis XIV as the ten-year-old King of France and Navarre. He is dressed traditionally (and rather simply) in a style closely related to the medieval kingship. In similar portraits later in life, one sees less Louis—and much, much more regalia.

Exercises

1. Define the following terms:
   a. System.
   b. Environment.
   c. System life cycle.
   d. Stages.
   e. Failure.
   f. Success.
   g. System roles.
   h. User.

For each of problems 2 through 6, identify the initiation and operation stages:

2. Shelter.
3. Automobile.
4. Payroll system.
5. A new medicine for polio.
6. Mars probe.

Identify the four roles involved in the following:

7. Erecting a government building.
8. Building a private residence.
9. Taking the family out to dinner.
10. Planning a vacation.
11. Establishing zoning along a coastal area.

What stage applies to each of the systems in problems 12 through 18? Why?

12. You are adding a room to your house and have just had the building plans approved.
13. You are thinking about going out to dinner. You find that you have $3.00.
14. A child has $.25 and is trying to decide which candy bar to buy.
15. You decide that you will change the oil in your car yourself.
16. You are selecting a chair for the living room in your apartment.
17. A new government agency is arranging its staff on the basis of an approved organization chart.
18. You have just bought a garbage disposal. It was a good deal. You are now reading the instructions for putting it in.

In each of problems 19 through 22, explain in systems terms what could have been done to prevent failure.

19. Some time ago, hydrogen-filled dirigibles were popular. The *Hindenburg* was one such airship. It exploded on arrival at Lakehurst, New Jersey, while attempting to dock.
20. The army had planned for a new major tank. After all of the design work had been completed, it was found that the tank was wider than the roadway of many bridges in Europe. The design was dropped.
21. The Aswan Dam across the Nile was built to provide power to Egypt. However, due to the tropical setting there is a substantial loss of water from evaporation. There is also a depletion of soil because the Nile floods have ceased.
22. A study of compulsive gambling behavior was funded for a twenty-six-week period in Las Vegas. A program was designed based on ten consecutive interviews with the subjects. It failed because the average hotel stay in Las Vegas is four days, which does not allow implementation of the sequential questionnaires.

In each of problems 23 through 26, the outcome was more successful than originally anticipated. What could have been done to predict success?

23. In 1948 Harry Truman won a totally unexpected victory over Thomas Dewey in the United States presidential election. The polls had predicted Dewey's victory by a landslide.
24. The peasants of southern France had a folk remedy for infected open wounds—they applied Roquefort cheese to the infected area. Doctors scoffed at the peasants' superstition. What could a cheese riddled with mold do to help infection? (The discovery that penicillin is produced from such mold was still years in the future.)
25. After World War II, many people who knew about television thought it would be unsuccessful due to expense and technical problems.
26. In the early days, crude oil was refined to produce only lubricants; its by-products were discarded as waste. Today technology has developed and is continuing to develop other uses for these by-products. Among these are polyethylene, polystyrene, synthetic fibers—and, of course, gasoline.

NOUVEAU PLAN des VILLE, CHATEAU et JARDINS de VERSAILLES

Dessiné sur le lieux en 1714, avec la marche que le Roy a ordonnée pour faire voir le Jardin, les Bosquets et les fontaines du dit Chateau Royal de Versailles.

le Canal

la Piece des Suisses

rue de la Surintendance

rue des Reservoirs

L'Estang

Avant Cour

Place Dauphine

Parc aux Cerfs

le Marché

2

System Environments and Roles

We started off this examination of the systems approach with a close-up look at systems themselves: how they are born, live, and die, and what their components are. In this chapter, we'll see how they exist in their surroundings. We will zero in on the complex interfaces that allow a system to operate.

As the scientific world has so usefully taught us, nothing on earth exists in a vacuum; the laws of motion do not apply without the mitigating forces that exist in any physical environment. The same is true of systems. Perhaps the biggest failing of many systems analyses is the analysts' reluctance to look at the larger picture: the world in which the systems must exist and operate. That world, or *environment,* is in a constant stage of change; nothing stands still, even during the development of what might appear to be a simple, static system.

The Environment

The environment in which any system is created and must exist is composed of two overlapping groups of factors which are frequently in flux. First, there's the preexisting environment, composed of the factors that induce the creation of the system. Second, there's the combination of environmental factors that must sustain the system during its development and operational life. That is to say, it's the environment at large that contains the original problem requiring a systems solution, and it's the environment again that must contain and sustain the system during its useful life.

Generally speaking, there's considerable overlapping between the two groups of factors, and care must be taken to distinguish one from the other. *The overall aim of a successful system is to convert the first group of "problematic" factors into the second group of "sustaining" factors.* We'll discuss this at length during the rest of this chapter and throughout the remainder of this text.

No system can operate independent of its environment, because all systems are designed in response to an environmental need. The degree to which a problem solver neglects the problematic environment is the degree to which that problem solver risks creating a system that will fail.

We can usefully define the major groupings of environmental forces and conditions as follows:

GROUP I: The problematic environment: The factors we define as group I have created the problem and have given rise to the idea that a systems solution is possible.

GROUP II: The sustaining environment: This environment is composed of the resources (group II factors) that the operating system must draw on for life support.

This two-part view of a system environment is an essential component of our overall systems approach. We'll use this delineation throughout the text in referring to the factors that induce the creation of a system, and the factors that sustain it. Although the terms (*group I* and *group II*) need not carry forward into future system development that the reader may participate in, the concepts must. We've used these terms uniformly throughout this text, and the reader should become familiar with them before reading further.

Each of these environments—problematic and sustaining—is made up of *forces* (factors in a state of change or motion) and *conditions* (factors at rest or static). The successful system must address itself to all the factors in both major groupings.

During the previous chapter, we discussed Thomas Edison. He lived and worked within the boundaries of a restrictive problematic environment, and he tailored his problem solving to the environment in which his solution would perforce live or perish. Figure 2.1 shows some of the forces and conditions that were active in Edison's situation.

FIGURE 2.1
_____

Partial List of Environmental Factors Pertinent to Edison

GROUP I

1. The spread of education, which created a need for better indoor lighting for reading.
2. The incidence of fires caused by gaslights and whale oil lamps.
3. The structure of the great industrial concerns, which excluded Edison.
4. Underutilization of the cheap energy of electricity.

GROUP II

1. The growing consumer market.
2. The availability of materials.
3. The Industrial Revolution, which encouraged the development of new machinery for profit.
4. The state of knowledge and development of the sciences, which provided the tools for research.

While figure 2.1 does not purport to be a complete analysis of Edison's case, it gives a good view of the types of factors active in his situation and of the considerable overlapping between the group I and group II components.

Developing a Group I List

Development of a list of group I factors is one of the first essential tasks a problem solver must face when confronted with a need for a new system. Each student will probably find a method of developing such a list suited to his or her own analytical technique. We suggest the following method:

1. Without regard to the concepts of forces and conditions, make an exhaustive list of the factors present at system inception.

2. Once the list appears to be complete, analyze the factors listed and decide which are dynamic and which are static.

3. Divide the list on the basis of forces and conditions (dynamic and static).

It's unusual—perhaps impossible—for one factor to appear in both columns (force and condition), but the student should find distinct corollaries between the two.

A careful analysis of figure 2.1 reveals that in the group I column factors 1 and 3 are conditions; each describes a static element present at system inception. Factors 2 and 4 are forces.

Group I forces are the ensemble of demands that require the problem solver to create a new system. Group I conditions are the broader, more diffuse set of factors that describe the environment that contains the need for a new system. We will deal at length with the definition and uses of a group I list in chapters 3 and 4.

Conversion of Group I to Group II

Having made up a group I list, which describes fully the components of the problematic environment, we are ready to begin developing the system solution. A *system solution may be defined as that set of actions, guidelines, structures, and/or procedures that best satisfies the need contained in the problematic environment by the most efficient use of the resources contained in the sustaining environment.*

The main task here is to perceive the elements of the sustaining environment (group II) within the problematic environment (group I) and to create a system that efficiently reshapes the group I factors into group II factors. Ah! But you say, "Isn't the main function of a problem solver to create a system solution to the problem?" Of course it is! We're simply restating that most obvious function in terms that are more easily usable.

Anyone can create a system; all of us do create them every day. The value of systems lies not in the fact that we can create them, but in our ability to create them to be effective—to be successful. And in order to be successful, they must be in harmony with their environments. Remember that failure is the prevention of utility; the corollary is that success is the facilitation of utility, and it is equally true.

Anyone can create a system; only the careful problem solver can create a system and ensure its success. And even the careful problem solver can ensure success only to the extent that he perceives and uses his environmental resources.

We will delve into the all-important conversion of group I to group II factors throughout chapters 5 through 9. For the moment, let's return to the village of Versailles for an initial look at the overall environment in which the Versailles system was created.

The Environment at Versailles

During Louis's reign, France is the most powerful nation in the Western world—the single greatest seat of influence from the Volga River in Russia to the coastline of California. The environment in which Louis operates is breathtakingly large in scope. It requires solutions to problems that can be perceived as important halfway around the world.

The problems Louis faces are equally staggering (a circumstance that seems to plague world powers of all ages), and the environmental influences range in scope from Scottish Presbyterianism and French silk manufacturing to English piracy in the Spanish West Indies.

The problems that require the most urgent attention, however, are those closest to home. Enter the Fronde.

The Fronde

The Fronde was made up of the great and powerful families of nobility who traditionally vied with the French monarch for control of France. It was actually the first development of the event we generally recognize as the French Revolution. A general uprising against the monarchy itself, it took place while Louis XIV was still officially underage.

The Fronde was a system that failed miserably. It was an ill-structured system with no central coordination and no common purpose. It totally ignored the problematic environment and found itself fighting not only the monarchy, but also the people of France and—more important—the entrenched bureaucracy of the courts.

Because it failed, the Fronde is largely forgotten by history. But it wasn't forgotten by Louis; it became the single most important group I force in impelling Louis toward his system solution at Versailles.

Louis accepts the submission of the nobility after the demise of the Fronde. Notice the "new look" in Louis's appearance. Soon these simple aristocrats will be flouncing about in ribbons and laces as well.

The dynamic of the Fronde was based on a group I condition that might best be described as the independent power of the nobility. The thrust of the Versailles solution is to convert that energy to the support of the monarchy by eliminating the independence that had previously encouraged the nobles to rise against the monarchy. By building and implementing Versailles, Louis will manage to make the nobility dependent on him for prestige and income and convert the untamed energy of the nobility to the mainstay of the monarchy.

Sound complicated? It really isn't. What Louis will do is to move the powerful nobles out of their homes in Paris and force them to live in his home at Versailles. There they'll have to live by his rules and his schedule; they'll be given no incentive to rebel, because the best of everything will be theirs for the asking, *as long as they behave.* By 1680, Louis will have only to utter the sentence, "He is a man whom I do not see," to condemn even the most influential member of his court to absolute social exile. To be sent away from Versailles—the most magnificent court the world had ever seen—will be to be cast out into darkness. And anyone who falls under the spell of Versailles automatically will become Louis's puppet.

So Versailles will become Louis's insurance policy against another occurrence of the Fronde. He will successfully convert all the group I components of the Fronde to group II resources, which he can use as he sees fit.

The Economic Situation

In addition to the considerable woes of the Fronde, France is stagnating in an economic depression, the result of numerous civil crises that have racked the state: Catholic versus Huguenot, monarch versus noble, France versus Spain, foreign trade versus domestic industry.

The royal treasury is depleted, and taxes are ruinous. French industry is dying on the vine for lack of innovation and markets. The French luxury industry (which we see today as perhaps the largest staple of French export—wines, perfumes, textiles, and furniture) is nonexistent. The great wealth of the Western world is concentrated in the mass of little states of Italy.

For our purposes, the population of France at the time of Louis XIV's accession can be divided into four main parties. Figure 2.2 sketches some of the most important aspects of these four opponents. All the forces and conditions that appear in figure 2.2—together with myriad others—make up the problematic environment, or group I factors, that Louis faces during his reign.

Perhaps the most persistent French exports are food and wine. These period etchings of farm laborers illustrate the role of the analysis and development group in the countryside.

FIGURE 2.2

Organizations in France During the Reign of Louis XIV (1643–1715)

	MONARCHY (LOUIS XIV)	PEOPLE OF FRANCE	BUREAUCRACY	NOBILITY
COMPONENTS	1. Royal family and dependents 2. Ministers and dependents 3. Most clergy	1. Citizens of Paris 2. Bourgeoisie 3. Other French citizens	Parliaments (courts) and their officers (presidents)	1. Great families of nobility 2. Nobles of clergy
ALLIANCES	1. Foreign nations 2. People of France	Foreign merchants	None	Minimal
FOREIGN INFLUENCES	Situation in England (beheading of Charles I)	Papacy	Situation in England (revolt of Parliament under Cromwell)	None (noble clergy fiercely Gallican)
ANTAGONISTIC TOWARD	Nobility	Nobility	1. Monarchy (at times) 2. Nobility	Monarchy
REVENUE STATUS	Levies taxes	Pays taxes	Partially tax-exempt; pays import taxes	Tax-exempt
ECONOMIC STATUS:				
START OF REIGN	Treasury depleted	Poor	Wealthy, independent	Wealthy, independent
END OF REIGN	Treasury depleted	Poor	Wealthy, partially independent	Wealthy, dependent

The decision to build the château at Versailles is based on Louis's careful analysis of what we are calling group I factors. His remarkable management of France during his reign will so successfully convert enough of these group I factors to group II factors that his system of government will outlive him by three generations. If he would have even more carefully converted these factors, more completely converted the problematic environment, France might well still be living under Bourbon rule in the late twentieth century. As it is, Louis is a man of limited foresight and does not actually operate under the precise systems approach we attribute to him.

The Four Roles at Versailles

In chapter 1, we defined four roles active in the development of systems: user, management, project leadership, and analysis and development. The relationships of these four roles to each other are diagramed in figure 2.3. In a business situation, the user might frequently be called a *client;* management functions as *middle and upper management* (line and staff); project leadership functions as the lower of *line management;* analysis and development assumes the role of the *work force.*

FIGURE 2.3

Interfaces and Reporting Order of Four Roles

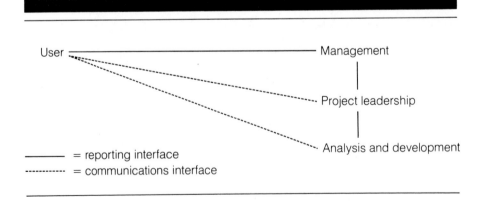

User — Management

Project leadership

Analysis and development

——— = reporting interface
----------- = communications interface

The flow of authority and information depicted in figure 2.3 is closely comparable to the situation that we may imagine is created by Louis XIV in building Versailles. Louis himself plays the role of user; his chief financial minister, Jean Baptiste Colbert, functions as management; the architects and designers of Versailles represent project leadership; the skilled craftsmen, artisans, and laborers who build the château can be designated analysis and development. As will frequently be the case in the development of systems in the future, Louis as the user never adopts a hands-off policy in regard to Colbert's (management's) authority to regulate the project. He will continue to dictate requirements throughout his seventy-two-year reign.

The User: Louis XIV

We'll consider that Louis functions as a typical user in our case study of Versailles. First, and most important, he discovered the problem: his kingdom was unhealthy, explosive, and disunited, and the power and prestige of his monarchy were dwindling.

With only a vague idea of a solution in mind, he persuades Colbert to take on the task of management. Colbert is appointed to a position equivalent to a modern commissioner of public works, from which he manages the king's building projects. Then, working with Colbert and Colbert's subordinates, Louis ordains the masterful shift in French politics represented by the decision to build Versailles: he decides to move the whole power base of the French government out of the hands of all potential opponents—out of Paris and onto the insignificant estate at the village of Versailles.

Having discovered the problem and worked closely with the "project team" in the development of a solution plan, he provides the major requirements throughout the life cycle of the project. Few of the major requirements dictated by Louis are based on personal whim; the château is designed to symbolize and effect the rebirth of the French nation. And it will work.

Management: Colbert

Within the parameters set by Louis's requirements, Colbert functions as a model manager. He oversees the essential conversion of group I factors to group II resources; he provides rock-solid financial and line management; he makes all important management decisions and oversees all major construction.

Unfortunately, he won't live to see the château in the form in which Louis will finally leave it; he'll die in 1683, a full thirty-two years earlier than Louis. Nevertheless, we shall for the purposes of this case study consider Colbert the management role throughout the project, because his methods and influence will outlive him by as many years as the construction of the château requires. His successor will follow the same project plan with the same administrative style, for all practical purposes. Colbert will outlive himself through the organization he set up and the men he chose as his successors.

Jean-Baptiste Colbert: project leader par excellance *of the construction of Versailles.*

Project Leadership: Architects, Designers, Artists

The seventeenth century is a period of great achievement in the history of French genius. Much of this achievement is a direct result of Louis's decision to build Versailles. The most lasting effect of the construction of the château will take place through the influence and genius of these men who were Versailles's project leaders: they'll reshape the economy of France. Ever since Versailles, France has been perceived by the rest of the world as the center of fashion, sophistication, luxury, and taste.

It's important to understand that for two hundred years prior to the construction of Versailles, Italy has set the pace in style and taste. The splendor of the Medici, the Borgia, and the Sforza families has captured the imagination of Europe. No artistic or mercantile idea of importance has emerged from France in nearly four hundred years.

Today's French luxury industry needs no introduction. Throughout the world, Rue de la Paix means perfume; Paris means designer clothing; and Bordeaux means wine.

From the point of view of the people involved, the impact of Versailles on French industry will be epochal: France will change almost overnight from an importing country to a major exporter of manufactured goods. And it will maintain that status to our own time.

Analysis and Development: The People Who Built Versailles

Thousands—possibly hundreds of thousands—will work on Versailles during the reign of the Sun King. No project in the history of France has ever equaled it. Few projects in the history of the world rival it.

Surveyors, seamstresses, political philosophers, judges, winemakers, actors, mistresses, carvers, farmers, ditchdiggers, draftsmen, bricklayers, stonecutters, gardeners, and gondoliers—all these and hundreds more will make their contributions to the new center of the world (as it is intended to be). Each will report to a project leader, and each will tailor his contribution to the specific requirements of the ubiquitous user: Louis. No artisan who will work on the château will work independent of Louis: Louis will oversee every detail of importance. And every detail is important in the grand system that Versailles represents.

A vision of the New France—vigorous, industrializing, and a bit bizarre. Where else could a portrait of a blacksmith be done in the general shape of a peacock?

The Roles Together: A Pyramid with the Sun at Its Peak

The four roles at Versailles work much like a system in a late twentieth-century setting. The specific relationship of roles and decisions is represented by figure 2.4. The shape of the organization is dictated by the need of Louis as user to focus the attention of the world on the French monarchy. The shape

FIGURE 2.4

The Four Roles at Versailles

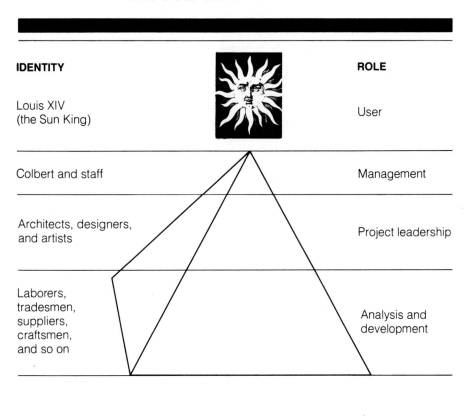

IDENTITY		ROLE
Louis XIV (the Sun King)		User
Colbert and staff		Management
Architects, designers, and artists		Project leadership
Laborers, tradesmen, suppliers, craftsmen, and so on		Analysis and development

of the organization itself accomplishes part of Louis's system solution: everyone in the pyramid structure now depends on Louis for personal sustenance, just as Louis depends on the mighty structure itself to carry out the solution dictated by the problematic environment.

Most modern systems aren't able to function in this precise pyramid pattern due to environmental forces that exert distorting influences on the structure. To some extent, however, most systems organizations that report to a strong management section carry important elements of the pyramid structure pictured in figure 2.4.

Management Organization: Types of Decisions

Any management organization is faced with many different types of decisions in each subject field it deals with. A manager in a modern corporation makes decisions in three basic time spans:

1. Day-to-day, which covers the decisions that require immediate action to effect the short-range outcome of a project. These decisions are called *tactical decisions* and are made by a management subset (*tactical management*).

2. Medium-range, the span into which fall the decisions that require planning and projections of six months' to one year's duration. Yearly budget projections are examples of medium-range decisions. We'll refer to this area as *managerial decisions*.

3. Long-range, which encompasses decisions that require projections and planning of more than a year's duration. These are called *strategic decisions* and are made by *strategic management* as the dominant subset of the management organization.

Looking at these three levels of decision making from a different perspective, we can say that tactical (day-to-day) decisions are made entirely on the basis of internal information—information which comes entirely from within the organization. Strategic (long-range) decisions are made largely on the basis of external information: information generated outside the organization. Managerial (medium-range) decisions contain elements of both internal and external information.[1]

From this perspective, we can see that marketing decisions are largely strategic, because they deal largely with external information (the outside marketplace). Employee-related decisions are managerial, because they deal with internal information in light of standards prevalent outside the organization. Product manufacturing methods involve tactical decision making, because they are almost totally internal to the particular organization which manufactures the product. Other examples can be generated using the same criteria.

In our case study at Versailles, Colbert and his staff have to make decisions in all three categories. Colbert's project leaders may be faced with tactical and

---

1. Either set of criteria may be used. Decisions may be categorized on the basis of time span or sources of information. The two will usually coincide.

managerial decisions delegated to them by Colbert, but no strategic decisions are made outside Colbert's circle of immediate control.

It's essential to provide mechanisms for making these three types of decisions in all management organizations. If responsibilities for these three ranges are not separated carefully, the sheer quantity and burden of tactical decision making may tend to squeeze out the difficult strategic decision making required in long-range planning.

Summary: what do we know now?

We've examined the problematic environment that engulfed Louis XIV, and we've categorized the forces and conditions that make up the problematic environment as group I factors. We've seen that Louis and Colbert must successfully convert these group I factors to group II factors (the sustaining environment) in order to make the Versailles system work. We've acknowledged the environment as the most important component of systems.

We've identified the four roles at Versailles and discussed some of the lasting impacts of the system they created. We've examined the interfaces and reporting relationships that allowed the project team to function, and we've identified three types of decision making: tactical, managerial, and strategic. And we've examined the dominance of strategic decision making in a well-structured organization.

Where are we going?

In the next chapter, we'll begin a detailed examination of a hypothetical life cycle at Versailles. We'll look critically at the systems approach as it applies to our case study—and we'll see how things might have happened if the principles of systems analysis had been consciously applied.[2]

And we'll build our palace.

---

2. Which, of course, they were not. The systems approach was not a twinkle in Louis's eye. We are not attempting to change the outcome of Louis's project—nor to propose a new view of history. Our Versailles is in many ways a fantasy, a piece of historical science fiction.

Exercises

1. Define the following terms:
   a. Group I.
   b. Group II.
   c. Forces.
   d. Conditions.
   e. Tactical.
   f. Managerial.
   g. Strategic.

In each of problems 2 through 5, a situation is described together with a list of factors. Classify each factor as to whether it is a force or condition. Give reasons for your answers.

2. In the early twentieth century, Henry Ford devised the idea of a production line for automotive production.
   a. Demand for cheap cars.
   b. Low-cost labor.
   c. Industrial Revolution.
   d. Profit motive.
3. A boy wishes to adopt a dog. He finds out that he can get one from the city animal shelter for five dollars. He buys one and brings it home.
   a. Small amount of money available.
   b. Desire to have something to care for.
   c. Parents' resistance to getting a dog.
4. The Colosseum in Rome was built as a public works project to entertain the people and to enhance the image of Caesar.
   a. High unemployment.
   b. Military acquisition of wealth.
   c. Available land.
5. A labor negotiation session breaks down due to a disagreement over benefits. A strike begins.
   a. Rising medical costs.
   b. Federal tax rates.
   c. Inflation.

In problems 6 through 9, a series of situations is described. In each case compile a group I list; divide it into forces and conditions, and describe the method by which you would expect them to be transformed into group II factors. Your list need not be exhaustive or complete; simply select the most important forces and conditions for your discussion.

6. A successful civil lawyer decides to run for a Senate seat in the upcoming election. He is not wealthy; he has a well-known party affiliation; he has never held elective office before. His state holds Senate primary elections.
7. A real estate developer wishes to subdivide and improve a parcel of coastal land protected by a sluggish coastal commission. The developer is willing to allocate a large percentage of his acreage to greenbelting and to stay within the height restrictions imposed by the commission. He must, however, obtain permission from the commission before he can float a bank loan. The commission requires a thorough environmental impact report with the application; such a report will cost $50,000—a sum that the developer does not have available.
8. A nationwide merchandising company finds that it is losing large amounts of money in its credit card operation. The company decides that elimination of the credit system is the best alternative, together with the acceptance of national general purpose cards. They suspect, however, that such a step will seriously damage their national brand status and will anger their regular credit customers.
9. A hotel chain is committed to building a new hotel in a country where currency is gaining value in relation to the American dollar. The hotel chain, an American company, wishes to avoid the ever-increasing expenditure of dollars that would be required to keep pace with the host country's currency. Social unrest in the host country threatens to disrupt the government there, and so makes the long-range future of its currency unsure.

In each of problems 10 through 13, a system at least partially failed. Identify possible sources of failure by isolating group I forces that were not transformed into group II conditions or the group II forces that the system failed to handle.

10. In 1917 the Russian government fell, and a limited representative government was established. It was later overthrown by the Bolsheviks.
11. The *Hindenburg* was a hydrogen-filled airship intended to provide another means of transportation across the Atlantic Ocean. The airship's hydrogen exploded during a landing at Lakehurst, New Jersey.
12. Food stamps were intended to be a short-term low-cost way of alleviating hunger and using up the agricultural surplus in the 1960s. They now constitute a multi-billion-dollar program, which is under attack.
13. The BART transportation system was intended to replace automobile traffic in the San Francisco traffic corridors connected to Oakland. The company that made the cars for the rail system has ceased to make them. Control problems occurred. Ridership is below that expected.

In problems 14 through 17, the system was at least partially successful. Explain what group I forces were transformed and how the system transformed the forces.

14. The interstate highway system is now over 70 percent complete. It provides a marginal cost means of individual transportation.
15. After the start of World War II in 1939, United States unemployment ceased to be a major problem.
16. The "Sesame Street" show on public television succeeded beyond expectations.
17. The Apollo spacecraft landed successfully on the moon.
18. Review the cases summarized in problems 6 through 9 and your analyses of the group I and group II factors. Identify areas of decision making as tactical, managerial, or strategic. Justify your decisions with reference to time and outside information involved.

In each of problems 19 through 23, discuss what internal and external information is needed. Suggest possible sources for the information.

19. The decision to build a fast food outlet in a town.
20. The problem of deciding on the type of marketing campaign to reduce electricity usage.
21. The decision to invest in a politically volatile country to extract minerals over a five-year period.
22. The decision to run for City Council.
23. The decision as to which archaeological site to explore.

NOUVEAU PLAN des VILLE, CHATEAU et JARDINS de VERSAILLES

Dessiné sur les lieux en 1714, avec la marche que le Roy a ordonnée pour faire voir le Jardin, les Bosquets et les fontaines du dit Chateau Royal de Versailles.

3

The Initiation Stage

In the first chapter, we examined the concept of a system life cycle from a broad vantage point. Beginning with this chapter, we will examine that cycle stage by stage, close up. We'll examine the types of decisions required to keep a project going. We'll explore the problems of scheduling, budgeting, and personnel. And we'll begin to build a grid to organize projects from the highest strategic decision-making perspective down to the most nitty-gritty operations.

Consider now that we have acquired a working knowledge of the flow of systems through the life cycle: initiation, feasibility, analysis, design, construction, installation, and operation. We have decided that one of the chief assets of the systems approach is the ability to commit resources by stage and not by project. That asset alone has been enough to convince many budget-conscious organizations of its worth.

It's a peculiar strength of the systems approach that it isn't an all-or-nothing method. It's a way of thinking that encourages—indeed, demands—constant reevaluation of objectives and resources. It facilitates the type of decision making that enables a project to achieve maximum success. But with that bit of philosophizing out of the way, here we sit with no project yet started.

41

The Guy with the Itch

As we pointed out earlier, there are two functions in the creation of a system: *system development* and *use*. It's the second of these, use, that gets the ball rolling. The user is the person with the need for a system; he needs it to solve a particular problem or set of problems. Actually, although we call him the user, he may not be using anything just yet; so, at this phase, we might just as well call him the needer.

This user/needer usually has some perspective on the problem—perspective that may be gained in a negative fashion (he's had a history of problems), or in a positive fashion (he's got a brainstorm), or in a combination of the two (he's got a problem and solution, but no methods). But he does have more ideas on the subject at this point than anyone else, so he is the first source of information for the system development function.

In most situations, the user is an easy fellow to identify: he's the one with the itch, the one who needs to get something done. There are situations, however, when the user is considerably removed from the project; "he" may even be a nebulous group of people who have no direct connection with what's being done at all.

Often you are the user. You need something, and you get it done. But you wear two hats, and the two functions indicated by those two hats strike a bargain within you. One says, "This is what I need"; the other says, "This is what I can supply"; and you make a decision based on that information. It's a systems approach reduced to the simplest possible scale.

That's what Edison did. That's what made him a remarkable entrepreneur as well as a remarkable inventor. Not every inventor is—not even every important inventor. Marconi invented the radio, but it took David Sarnoff to start RCA (with Marconi standing at his left hand for the sake of public relations).

The Origin of Systems: The User

It's a very rare occurrence indeed when a user approaches a system development group for a solution to a problem that has no solution whatever in current force. Most users are already coping with the problem situation in some fashion, usually with an outmoded system, or a system that has outgrown its design. There are too few novelties in nature or society to require commonplace systems that are totally innovative. So a first law of system development might be stated: *The problematic environment that requires a system solution already contains a solution, although that solution is unsatisfactory.* Included in this category might be a dam with unsatisfactory sluicing channels, a payroll system that has become a bureaucratic nightmare because of paperwork, or a political system that has become disoriented but is still functioning after a fashion.

It's precisely the latter that Louis XIV has to deal with. There's no doubt that Louis's France is functioning: people are living, eating, procreating, and working. The government is operative: the courts try cases; the tax collectors collect taxes; diplomats and princes go on foreign missions; the king and his retinue make day-to-day decisions when they're called for. Cardinal Mazarin (Louis's stepfather because of his secret marriage to Louis's mother, Anne of Austria) holds the reins, but only in the king's name.

But there is a nagging and very real problem in Louis's France. There is no sense of stability or continuity in the nation. Merchants don't take risks in trade. Courts defer too freely to hereditary privilege. Tax collectors cheat a little too openly. Nobles do as they please a little too freely. The nation is without central focus, and there is considerable jockeying for control; the situation is very unsettling to a monarch whose mind is fixed on the divine right of kings. He doesn't like relinquishing *anything* that touches on regal control to *anybody*. He'll later demonstrate this even in his choice of ministers; Mazarin will be the last minister in Louis's reign with high social as well as high administrative position.

The knowledge of French history is unimportant here; what is important is to realize that Louis has a reason for seeking a major change in his realm. That reason, succinctly, is a need for power: regal, personal, totally unquestioned power.

The Predicament of the User

Leaving Louis to stew, for the moment, in his own juices, let's consider the predicament of the modern user. He has a problem that he needs solved. But he doesn't have a solution, and he doesn't have a system development crew to hand it to.

The first decision to be made in the development of a system is the crucial user decision: *who to seek help from.* Obviously, with the breadth of problems that are susceptible to system solutions, we aren't able here to give guidelines for the selection of a system development team. But potential users should be aware that this decision is one of the most crucial ones in the entire course of system problem solving.

When there's a choice to be made, the user would do well to consider the alternatives carefully. The most common choice is between a "quick and dirty" solution in a temporary mode and a more expensive, long-range solution that does very little to ameliorate the present situation. Frequently, the choice of system development group dictates this sort of direction.

A common situation where this decision is crucial occurs in the real world. Take any problem department in a corporation; two alternatives can be pursued in solving the problems:

1. A personnel solution, which involves eliminating, retraining, restaffing, or reorganizing the personnel involved in the hope that next time something better will happen.

2. A system solution which requires initiating a thorough study of the situation, and acting on it in a deliberate, careful (and perhaps *slower*) manner.

Different situations will demand different answers to this decision. Acute crises will demand immediate answers; chronic problems are fruit ripe for plucking by systems organizations.

Let's presume that the user establishes some viable criteria for selecting a system development group to trust. In Louis's case, he found a single man he could trust: Jean Baptiste Colbert. He entrusted his system solution to Colbert, and Colbert established a system team that was remarkably successful.

Louis and Colbert: The Scope of the Problem

The problems that face Louis and Colbert (for now they had found each other and formed a strong alliance) are ancient. In a Europe that has grown up on feudalism and primitivistic trade practices, any attempt at revivification is doomed to failure almost from the start. Louis's kingdom includes a privileged nobility that may number one hundred thousand persons. Its history is ancient even in the seventeenth century, and many problems predate Louis's family's possession of the French throne (the dynasty is only sixty years old as Versailles is begun).

Although his position and prestige are never challenged after the defeat of the Fronde, Louis's family are Johnny-come-latelies to the throne of France. Louis's family had once been the kings of Navarre (unimportant enough that most people in three hundred years won't even know where it is) and for years were resident pensioners at the court of France because they were kicked out of Navarre.

The problem of the newness of Louis's dynasty can easily be overemphasized, but it's important. Louis has had to overcome the prestige of families whose histories in hereditary prestige are much older than his own. He could never uproot the aristocracy, and he probably will never want to. Who would kill a stable of championship racehorses? You don't kill animals like that; you might geld them, but you don't kill them.

Louis in procession. The original caption of this etching proclaimed that he "enchanted the eyes of all beholders."

While we leave Louis and Colbert to hash out the problems of seventeenth century France, let us turn to a task of the utmost importance in the initiation phase of system development: the compilation of a group I list.

Compilation of a Group I List

As we discussed in chapter 2, all problems that are susceptible to a system solution are composed of two environments: problematic and sustaining. Unfortunately, they don't appear in the sky laced with diamonds; they must be sought out and analyzed. And that's the function of a group I list.

We have divided environments into groups I and II in theory—and it's an easy thing to do in theory. It's not so easy in reality, but it is absolutely crucial to the success of your system solution. No solution exists or operates outside its environment, and no solution is successful if it flaunts its environment, runs counter to it, or fails to draw sustenance from it.

Consequently, from the very first word, *the key to success in systems is accurate, adequate, properly categorized information.* The first task that faces those who comprise the system development function is to interview the user or some representative of the user who's familiar with the situation.

Now, don't be shocked, but that interview presents a major problem in many cases because the user doesn't always know what the problem is. He has to know something is wrong (or at least not right), or he would not be consulting you. But he may not know just what it is that's wrong. He might be aware of a vague sensation of wasted time or money; he might be feeling pressure from a client because of inadequate product quality. But he may not be able to tell you what's wrong. So you get out your pen and start.

You begin with symptoms; he should be able to tell you these. You jot down the symptoms in the form of *environmental conditions.* Recall that we distinguished between *forces* and *conditions* in chapter 2; what we are after first is the latter.

The list of environmental conditions might include statements such as the following:

1. I'm not profitable.

2. My staff includes no young people.

3. My production line is outdated.

4. The lake is too large for the dam.

5. The highway won't carry enough cars.

6. The paychecks are always late.

7. My people don't keep me informed.

This list could go on ad infinitum (or ad nauseam). All of the above are environmental conditions that might apply to different problems. Remember that you must compile a list that's accurate, adequate, and properly categorized. In many situations, you may want to interview more than one representative of the user organization.

You'll be compiling, during the initiation stage (and to some extent during the feasibility stage), a list that will guide the development and maintenance of the system throughout its life cycle. There'll be plenty of chances later to revise it, but each revision is more costly than the last in terms of both time and money. Don't adopt a wait-and-see attitude about this list; get it all down now, using the best available knowledge.

Your list should be as complete as possible. It should include all the symptoms the user can name. Experience will breed a good technique for initiation interviews, but you must learn to draw from him the symptoms the user may have forgotten. Upon reviewing your list of symptoms or conditions, you may want to disregard some, which is easy. It's considerably more difficult to consider one that was left off the list—and, therefore, outside anyone's attention.

Creating Environmental Pairs

Once you have compiled a complete, accurate list of environmental conditions, you are ready to set about relating those conditions to *environmental forces*. The groupings of conditions to forces are termed *environmental pairs*. Figure 3.1 comments on the quality of the environmental conditions we've listed and pairs each with possible environmental forces. Each environmental pair should contain a condition that is caused or supported by a matching environmental force.

FIGURE 3.1

Environmental Conditions, Revised Conditions, and Possible Forces

ENVIRONMENTAL CONDITION	REVISED CONDITION	FORCE
1. I'm not profitable	Declining sales in an increasing marketplace	Increasing competition
2. My staff includes no young people	No new ideas	Hiring practices
3. My production line is outdated	Low productivity	Low availability of machinery
4. The lake is too large for the dam	Overflow of water	Rivers feeding lake are high
5. The highway won't carry enough cars	Traffic jams; accidents	Increasing commuter population
6. The paychecks are always late	High payroll procedures error rate	Small payroll department; staff is overworked
7. My people don't keep me informed	Lack of information flow	Poor reporting system

As we said before, the *environmental pair* is composed of a condition and a force that relates to the condition. Let's take an example: "I'm not profitable" is the first given condition. We can see immediately that it isn't precise. We need a "sharper" condition which pins down the situation. Examining the problem further, we find that the condition is really declining sales in an expanding marketplace. If we take this as our revised condition (see figure

FIGURE 3.2

Group I List: Possible Environmental Pairs

1. C = Declining sales in an increasing marketplace
   F = Increasing competition

2. C = No new ideas
   F = Hiring practices

3. C = Low productivity
   F = Low availability of machinery

4. C = Overflow of water
   F = Rivers feeding lake are high

5. C = Traffic jams; accidents
   F = Increasing commuter population

6. C = High payroll procedures error rate
   F = Small payroll department; staff is overworked

7. C = Lack of information flow
   F = Poor reporting system

3.1), we see that we must now develop a notion of a force. What is the force which is giving us a declining share of an increasing market? One force could be increasing competition. You should work through each of the entries to figure 3.2 to see how revised conditions and forces arise from the original condition. Remember, a condition is something that is stable. A force is dynamic.

The resulting list of environmental pairs is termed a *group I list.* It is the key to success or a condemnation to failure. All future efforts rise or fall on the quality of this list.

Louis and Colbert: Group I List

One suspects, looking at Louis's reign, that he is a fairly well-informed user. He knows where his problems are and he knows pretty much what he wants to do about them. He'll leave no records of any decision-making conversations between himself and Colbert. So we'll invent one. It might have gone like this. In interviewing Louis about the state of the kingdom, Colbert elicits Louis's gripes and complaints about France's symptoms and compiles a list of environmental conditions as follows:

1. Weak central control by monarch.

2. Bureaucratic chaos (primarily in the courts).

3. Walls created by hereditary privilege.

4. Low revenues to monarch.

5. Weak central government in general.

6. Flaccid economy.

7. Lack of innovation in trade.

8. Low international trade prestige.

You're right; it reads like a list of gripes. That's exactly what it is. And, within reasonable limits, it's fairly complete.

This list of environmental conditions tells Colbert and us that Louis's problems fall primarily into two categories: political problems and economic problems. But, sitting there in the middle of a page like that, the list doesn't give us any direction in which to move. For that, we need to find out what is causing the conditions. We need to create environmental pairs based on these conditions. To do that, we analyze each condition presented and determine which *environmental force* contains the energy supply that maintains it. Figure 3.3 is the group I list that Colbert presents to his client for approval. Louis wholeheartedly agrees with it. The items that appear in figure 3.3 will be discussed in detail in later portions of the text.

FIGURE 3.3

Louis XIV's Group I List

1. C: No control by monarch
   F: Powerful nobility

2. C: Bureaucratic chaos (primarily in courts)
   F: Judgeships hereditary

3. C: Hereditary privilege
   F: Tradition

4. C: Low revenue to monarch
   F: Impoverished and obstructionist taxpayers

5. C: Weak central government
   F: Louis's power base fragmented by recent history

6. C: Flaccid economy
   F: Civil disorder

7. C: Lack of innovation in trade
   F: Foreign trade is aggressive

8. C: Low French trade prestige
   F: No incentive to trade; no products to trade

Summary of the Problematic Environment Thus Far

The group I list compiled during the initiation stage is a summary of what we perceive in the problematic environment. It delineates our starting place in the search for a system solution to the user's problem. It will remain our most important checking device throughout the system life cycle; our continued observance of these original perceptions will guide us throughout the project toward success. The more we tailor our system solution to each of the components of the group I list, the more we're likely to create a successful system.

But, in order to progress any further, we need a direction and signposts to tell us which way to march in our search. The first task in setting up these signposts is the compilation of what we term a *tentative group II list*. It's our first stab at finding the sustaining environment hiding within all the problems. Remember, it's got to be there somewhere; when we've found it, we've taken a giant step toward a successful finish.

In our attempt to change the environmental conditions, which represent the symptomatic problems that the user is encountering, we'll first concentrate our analysis on the forces that support these conditions. These forces are the base source of energy, which we must convert and transform to nourishment for our system solution.

Compiling a Tentative Group II List

In this first pass at locating the sustaining environment, we will analyze closely the second members of each of the environmental pairs in our group I list: the problematic forces. For each force in the problematic environment we must formulate a tentative positive response. Never mind coordination just now. The first attempt must be at a sort of simpleminded utopia—whatever Pollyannaish response would succeed in the best of all possible worlds. Remember that this is just our first try, and we'll be back for further analysis several times.

In some cases, the items that appear on the tentative group II list may be simple restatements in positive terms of essentially negative forces. Consider the items that appear in figure 3.4.

Each of the responses listed in the tentative group II column is predicated precisely on the corresponding force in the group I list. And, while each of them may seem a bit simplistic, together they represent to Louis a list of objectives for his social, economic, and political reform of France.

To determine the means to accomplish the objectives stated in the tentative group II list, we must make our first stab at independent analysis of the problem. We will leave Louis in the Louvre and send Colbert out on a fact-finding trip.

FIGURE 3.4

Louis XIV—Group I List and Corresponding Tentative Group II List

GROUP I LIST	TENTATIVE GROUP II LIST
1. C: No control by monarch F: Powerful nobility	Reduce power of nobility
2. C: Bureaucratic chaos F: Hereditary judgeships	Reform court system
3. C: Hereditary privilege F: Tradition	Change tradition to support the monarch
4. C: Low revenue to monarch F: Impoverished and obstructionist taxpayers	Reform tax collection structure
5. C: Weak central government F: Louis's power base fragmented by recent history	Establish new, effective power base
6. C: Flaccid economy F: Civil disorder	Restore confidence
7. C: Lack of innovation in trade F: Foreign trade is aggressive	Make foreign goods unfashionable
8. C: Low French trade prestige F: No incentive to trade and no products to trade	Encourage French industry

Enter the Analysis and Development Role

The ball is now in the system development court. The initial interview of the user has been completed and a group I list has been compiled. Working together, system development and the user have agreed upon a tentative group II list which states the objectives of the system solution.

It's at this point in the system life cycle that the members of the analysis and development team begin their labors. Using whatever tools, resources, and information lie readily at hand, the A&D team must undertake a preliminary analysis of the problem and must then suggest a possible list of system solutions to the problem. This list of possible system solutions (or *alternative solutions*) will serve as the basis for the final user interview in this phase.

We visit libraries; we interview people who are (or might be) involved in the project; we read case histories of previous, similar projects. Standard research techniques are applied with insight and vigor, always under the guidelines set out by the group I and tentative group II lists. The research accomplished in this phase needn't be exhaustive in depth. If we decide to build a bridge to solve a transportation problem, we don't need to investigate the art of bridge engineering. We simply need to decide that a bridge is one of many possible solutions that would improve the flow of traffic across a river. If we decide to computerize a payroll system, we have no concern at this point for which software packages most closely resemble the likely solutions we may eventually need to develop.

What we must do is concentrate on the breadth of our research. If we are looking at a river transportation problem, we must consider tunnels, bridges, ferries, helicopters, land detours, and existing inadequate facilities. We wouldn't even be out of place looking briefly at broad jumpers and strong swimmers. We must be able to say with conviction that we've considered all the possible alternatives to the problem in our analysis.

There are two good reasons for not concentrating on the depth of our research at this point:

1. Only one alternative from our list will be chosen. Any in-depth research on the rest of them will have been wasted.

2. The initiation stage is usually constrained by a small budget and short time. The pressure is for a "quick and dirty" consideration of the problem. The only way to develop a satisfactory overview of solutions is to stay on the surface of each alternative.

Remember that one of the chief advantages of the systems approach is that it doesn't overcommit resources at any point. To undertake in-depth analyses of each of the alternatives supplied to the user at the end of this stage would be a violation of the charter given to system development by the user. He has

The automobile traveler approaching San Francisco can select one of several alternatives for entering the city. The Bay Bridge is seen in the foreground. The Sausalito Ferry plies the waters from the Embarcadero. The Golden Gate Bridge is in the distance. And, since the city is a peninsula, one can also travel southward and enter via dry land.

come to us for a systematic investigation—not for a harum-scarum research project that *by definition* must discard all but one alternative.

Colbert's Research

Colbert's analysis and development team (at this point a nucleus of thinkers and researchers only) will first have to analyze the problems presented in Louis's group I list (figure 3.3, page 51). Even a cursory glance shows us that the problems fall chiefly into two categories: political and economic. Items 1, 2, 3, and 5 are essentially political environmental pairs. Items 4, 6, 7, and 8 are blatantly economic woes. Consequently, any alternatives that Colbert proposes to Louis must tackle both monsters; we're sure from the beginning that no solely political solution will solve the whole problem.

The Colbert A&D workers must determine several basic facts. Among those facts are:

1. The numbers of people involved—nobility, bureaucracy, tradesmen, and peasant taxpayers. The answer must determine the size and scope of the problem. In Louis's case, the population runs to tens of millions; the nobility numbers tens of thousands.

2. The adaptability of current solutions. The answer to this category will dictate whether we must "start from scratch" or only institute a far-reaching reformation to solve the problem. In this case, there are institutions that will stand no tinkering: the monarchy, the privileged nobility, the long-established court system. None of these can be attacked frontally without invoking strong resistance.

3. Solutions to similar problems elsewhere. The history of Europe is full of monarchs who tried successfully or disastrously to solidify and strengthen their political and economic control. For Colbert, the most recent case, which is almost contemporary with Louis's problem, occurred in England. King Charles I tried to ward off the encroachments of Parliament and the nobility on what he considered his turf. Charles's approach was military and tyrannic. The outcome was Charles's eventual conviction for treason and execution by beheading—not a solution Louis wants to emulate.

4. The effectiveness of short-range solutions. Is there a patchwork solution? Can Louis, without disturbing the fabric of French society, effect a quick solution without a major revisionist project? The answer, in Louis's case, must be a resounding no; the problematic environment is much too widespread to be substantially affected by a bits-and-pieces reform. It must be an all-or-nothing solution.

5. Louis's strengths and capabilities. What are the king's resources—his particular strengths and traditional privileges? What is his fiscal position? To what extent can his coffers supply financing for change? The answer here is in two main sections:

(*a*) the French monarchy is in theory an absolutist government with unrestricted governing powers (although in practice, it is severely restricted by the encroaching privileges of the nobility); and (*b*) Louis is an extremely popular king during his youth—and an extremely effective politician. His financial resources are slight, but only because the tax base is weak; any substantial strengthening of the popular tax base will supply enough income for almost any undertaking.

With the answers to these topical research questions under his belt, Colbert is ready to compile a list of alternative solutions for his master.

Alternative Solutions

The compilation of a list of alternative solutions calls for a generous dose of creativity and insight. One of the alternatives proposed will be the germ of a project, and the hard work at this point must generate an idea that will create a viable solution. Needless to say, this is a crucial step.

The A&D team (which may consist, in smaller projects, of one part-time person) must sit down—pen in hand—and propose possible solutions. This amounts to a session of extremely well-informed brainstorming. But the A&D team must remain fixated on the list of objectives represented by the tentative group II list.

To illustrate the process, let's look again at the Edison example. Figure 3.5 suggests the range of alternatives that might have been considered in this case. Note that some of the alternatives seem almost fanciful, while some are foot-draggingly inadequate. The real, viable solution lies somewhere in between.

FIGURE 3.5

Possible Alternatives for Edison

1. Develop a more efficient way of extracting whale oil.
2. Concentrate on nonelectricity projects.
3. Use electrical direct connections without wires to connect homes.
4. Develop energy sources.
5. Devise new applications for natural gas.
6. Store sunlight for use at night.

The most fanciful alternative to any solution will contain some valuable insights or it would not be considered at all. If the Wright brothers had not envisioned men with wings, they would not have attempted to fly. Likewise, the most inadequate suggestions should bring to clearer focus some of the segments of the problematic environment that present the most difficult problems. The insights gained by fancy and by plodding are equally valuable, and even the alternative with the least plausibility should be included on the alternative list. After all, who would have thought that Edison could light a room with a tiny piece of wire?

Preparing for Feedback

Once a list of alternatives has been completed, the system development function must add to it whatever data, opinions, and experiences it can to point the way to the most probable solution. The picture presented must be full and accurate despite the fact that it's only a surface view.

What system development must concentrate on in presenting a list of alternatives to the user is the *range of those alternatives.* The value of the initiation stage is that it narrows the range of acceptable alternatives to a comparative few. And, while that narrowing is done in conference (system development and user functions both represented), it is done on the basis of the research that system development A&D has been able to collect.

Obviously, some alternatives developed during brainstorming should be eliminated before presenting the list to the user. But be circumspect in cutting—remember Edison.

Feedback

Feedback may well be the single most important component of the system approach. It's on the basis of feedback that all steps are taken, that modifications are approved, budgets allocated, and schedules drawn up.

Feedback is the response made to a flow of information. It's easy to define, no doubt about it. It's an easy concept. But, in practice, it isn't easy to maintain or to cope with. It's a process that can be carried on properly only between adults—that is, between people behaving in an adult manner. Watch a pair of children attempting to function on feedback someday. They'll eventually begin to fight or pout if they don't make their attempt in what we might term an adult mode. If they work together reasonably and coolly, they may reach a decision. The essence of acting in an adult mode is acting reasonably and coolly. Common violations of the adult mode in feedback are defensiveness, vindictiveness, emotionalism, haughty superiority, and groveling inferiority.

So much for who conducts feedback. Don't skim over the preceding, though, because you'll find yourself in situations where you must recite to yourself over and over that you will remain an adult at all costs. Feedback elicits that response from some people in some situations.

Systems thrive on healthy feedback of two varieties: positive and negative. Very few people have any trouble accepting positive feedback; it's reinforcing, encouraging, pat-on-the-back stuff. Negative feedback is exactly the opposite. It's commonly encountered as, "You're on the wrong track entirely," "After you fixed it, the faucet still leaked," or "You know that new program you wrote for the computer? It multiplied all the accounts payable records by ten!" While it may make the hair on your neck bristle (and let's admit it, it does sometimes), it's the most valuable kind of feedback we can encounter.

We'd probably keep moving along a determined course whether we got positive feedback or not. It's nice—and in some cases absolutely necessary—but it doesn't substantially alter the pattern in which things are operating. Negative feedback does alter that pattern. And, resistant to change as we all are, it sometimes grates, scrapes, and rubs the wrong way.

That "moving along/stopped in our tracks" dichotomy is what feedback is all about. The absence of feedback means that everything continues as is. Remember that you're not helping anyone by letting him or her spend more money and get in deeper just because you don't want to say something. Negative feedback is all-important. If you let someone travel the wrong road for only a few feet, it's an easy matter for him or her to return and restart; if you let him wander a hundred miles down a dead end, you won't be appreciated when he staggers back empty-handed from his quest.

So, feedback it is. The first feedback session in the system life cycle occurs when the system development function presents its list of alternatives to the user. Have a nice, soothing glass of warm milk, and plunge in.

The Culmination of the Initiation Stage

Colbert sits nervously in an antechamber of an antechamber of an antechamber in the Louvre. The king will see him now. He picks up his list and begins the ascent to the foot of the throne. That list he carries with him might well look like the one presented in figure 3.6, stating alternatives ranging from Louis's starting a foreign war, which will unify the nobility, to his retiring, as King Lear did, to his father's estate at Versailles, with all sorts of bureaucratic proposals in between—some of them valuable, some not.

FIGURE 3.6

Alternatives for Louis XIV

1. Start a foreign war to unify opposing groups.
2. Increase importation taxes on foreign goods.
3. Retire to family château at Versailles.
4. Issue wholesale edicts on change in the courts.
5. Increase taxes on selected groups.
6. Develop a new tax which can be collected by a less corrupt group of tax collectors.
7. Form a foreign alliance to overcome the power of the nobility and the courts.
8. Remove the seat of government from the control of the nobility.

From one vantage point, it's at just such a meeting that the destiny of France was probably decided. Although no concrete action has been taken, the decision Louis and Colbert make here determines:

1. The position of the nobility for the future one hundred years.

2. The functioning of French industry to the late twentieth century.

3. The fate of the monarchy.

4. The inevitability of the French Revolution.

5. The loss of French colonial possessions.

6. The magnificence of French baroque art.

7. The centralization of French government to the late twentieth century.

8. The intransigence of the French taxpayer to the late twentieth century.

That's a mouthful.

In actual fact, Louis makes all decisions of importance unassisted. He's that type of person. But Colbert is the man closest to the seat of power in all France. Later in life, Louis probably will tend to submit to the infamous Madame de Maintenon (his second wife) in political, international, religious, and economic questions; but in the period of his vitality, during which the fate of France is to be sealed, he is a decision-making loner.

Colbert supplies most of the information input needed for Louis to make decisions. Few others can get close enough to the king. Louis has never trusted the nobility enough to listen to them on political issues. He has appointed a long list of commoners to run his country; Colbert is the foremost of these commoners.

Today's system development function would operate on a much more equal plane with the user than Colbert did with Louis. But essentially the positions are the same. It is the user who must approve all major decisions, whose money is being spent, and whose veto must be unequivocably observed.

The culmination of the initiation stage is a meeting or series of meetings between system development and user functions, at which the list of proposed alternatives is examined and culled. The range of alternatives must be whittled down to a size that can be coped with. The focus of the project must be set, or feasibility cannot take place.

Louis and Colbert: The Decision to Move

It's at this point that Louis and Colbert make a momentous decision. After working through all the possible alternatives, they arrive at one that will deal with all the items on the group I list (figure 3.3, page 51) except one (the second—bureaucratic chaos/judgeships hereditary). It's a revolutionary alternative, considering the history of France. But it's a brilliant alternative—creative, innovative, original, and effective.

The decision is made to embark on the most ambitious public works project since the pyramids were built in pharaonic Egypt. This project will satisfy all the aching economic woes of import-weary France and stimulate French industry and ingenuity to one of the highest peaks in the history of man. It will produce a culture and artistic ideals which will still be respected three hundred years later, and make France an exporting country, rather than an exhausted importer of foreign goods. The nobility will be caught off guard; it will disembowel the aggressive former Frondeurs. It will tame the Great Ones and make glorified household servants of them.

It will make some extraordinary mistakes as well. With the implementation of this project will be planted the seeds of the harvest of sorrow known as the French Revolution.

Louis and Colbert decide to move the seat of government away from the Louvre and rebuild it in a locale where the entrenched nobility have no power. Many kings have built mighty fortresses and brilliant palaces. There is nothing out of the ordinary in Louis's decision to build a new house. The extraordinary decision is his determination to live in his new house exclusively—to govern from it, to promote from it, to distribute largesse from it, to entertain from it, and to make war from it.

Mad Ludwig, the nineteenth century king of Bavaria, built fairy-tale castles. Frederick the Great built a governing retreat at San-Souci, where he could cavort safe from the view of the public. The Roman emperor Diocletian built a titanic palace in Yugoslavia, which today has become the entire city of Split. George IV built the icy splendor of Buckingham Palace, where the honor and disgrace of the British Empire was heralded and hidden. But there is only one Versailles, and there has never been anyone remotely as grand as its maker, the Sun King, Louis XIV.

Onward and upward

With the selection of a general alternative ("move the seat of government"), we are ready to embark on the feasibility stage of the life cycle. In order to do so, we must obtain a commitment of resources and money to continue for another stage. Assuming that commitment to be made, we move boldly on.

Exercises

1. Define the following terms:
   a. Compilation of group I list.
   b. Group I list.
   c. Tentative group II list.
   d. Environmental conditions.
   e. Environmental pairs.
   f. Revised condition.
   g. Feedback.

In problems 2 through 4, the group I forces and conditions have been defined in error. Discuss what is wrong and propose new forces.

2. In 1973–1974, the United States and the other oil-consuming countries were subjected to oil embargos and oil shortages. The condition is the high demand for oil products in the United States. The force is the need to become self-sufficient in oil production.
3. Your neighbor has a teenage daughter who likes to throw wild parties when the parents are out. The condition is lack of parental control. The force is the parents' lack of concern.
4. You are a salesman for a company that makes and sells digital watches as well as traditional mechanical watches. The company has seen profit margins on all cheaper watches decline with competition. Your marketing budget is in jeopardy as a result. The condition is the increased competition and price pressure. The force is your retention of your marketing budget.

For problems 5 through 8, compile possible group I conditions and forces.

5. You run a coin-operated launderette in an urban area. There has been no competition for the last two years, since the old competitor closed. Now a new shopping center is opening six blocks from you. There is a new launderette planned.
6. You operate a small-boat launching ramp. People bring their boats on trailers and launch them from your ramp. The state government is studying the possibility of provid-

ing very low cost boat launching facilities near you. It could bankrupt you if the plan is implemented.
7. Your car is not running well; it needs a valve job and major overhaul. You just received an income tax refund of $1,500.
8. Recently in Mexico there has been a movement by landless peasants to seize agricultural land from absentee landlords.
9. For each of problems 5 through 8, compile a tentative group II list.

In each of the cases comprising problems 10 through 12, the system was preordained to failure because of a lack of effort and understanding in the initiation stage. Identify what factors should have been considered in each case to prevent failure. Explain in terms of conditions and forces.

10. The Edsel car.
11. The supersonic transport.
12. The dirigible *Hindenburg.*

In problems 13 through 16, study the situations as part of the initiation stage. Identify realistic information sources that you would use if you only had a few days to examine each problem.

13. The expansion of a real estate office to accommodate a 100 percent increase in staff.
14. The economic viability of a gas station.
15. The problem of busing in an urban environment.
16. The promotion of a movie.

In problems 17 through 20, discuss how failure could occur if improper group I and tentative group II lists were prepared.

17. A plan for using vacant office space downtown.
18. The problem of an overcrowded freeway system.
19. The lack of data on poisoning due to certain chemicals.
20. The potential need for a superluxury compact car.

NOUVEAU PLAN des VILLE, CHATEAU et JARDINS de VERSAILLES

Dessiné sur les lieux en 1714, avec la marche que le Roy a ordonnée pour faire voir le Jardin, les Bosquets et les fonteines du dit Chateau Royal de Versailles.

Le Canal

Logement des Matelots

la Piece des Suisses

L'Estang

rue de la Surintendance

rue des Reservoirs

Place d'Armes

Place Dauphine

Parc aux Cerfs

Le Marché

4

The Feasibility Stage

During the initiation stage, we discovered a problem that became severe enough to compel the user to seek help from a system development group. We did a lot of surface work to try to isolate the causes and symptoms of the problem, and we thought in a very vague way about what kinds of system solutions would best eliminate the user's itch.

At the end of the initiation stage, we still didn't have a solution. We spent the majority of our effort looking at what had happened in the past and at the problems the user was having. We compiled a group I list detailing the conditions and forces that defined the user's problem. We decided that the environment which contained the problem must also support the solution. Then we compiled a group I list detailing the conditions and forces that defined the user's problem.

During the feasibility stage, we'll expand that group I list (continuing our backward-looking work) and develop a firm plan for the future. Recall that, during our first discussion of the system life cycle, we identified feasibility with the fetal stage in human development. It is a very apt comparison. During feasibility, we must create an infant system that has the potential of full and useful life. It must be viable in its approach to the user's problem, and judicious in its consumption and replenishment of the user's resources.

Another Backward Glance

The first item on our feasibility agenda is a fuller consideration of the user's situation and of his problem. We'll accomplish this aim through standard research in libraries, data banks, periodicals, and whatever other information sources are available and relevant, and through interviews with persons concerned with the old nonfunctioning or malfunctioning system. Remember that virtually no problem exists that doesn't have some system (however makeshift) already coping with it.

Research techniques as such are outside the scope of this text. We won't tell you how to use a library, a computerized literature search, or a *Reader's Guide to Periodical Literature*. In addition, research techniques of this type vary widely with personal style and inclination. We will discuss interview techniques, however. They are critical to obtaining adequate and accurate information from which to project a system solution.

Interviews and Interview Techniques

An interview is a solicitation of information on a person-to-person basis. Another way to describe it is to say that it is a *request for feedback.* Since our growing system will be built largely on the basis of the information flow between the user and system development functions, we must take special care to ensure that that information flow is accurate and adequate. User interviews are the prime source of that ensurance.

There's a considerable amount of preparation needed for a user interview; probably more time is spent in preparation than in the actual interview. This preparation must be systematic and thorough. Basically, the preparation falls into two steps:

1. Develop any background materials you may need well in advance of the interview.

2. Prepare interview goals.

What are interview goals? Just what they seem to be. When you prepare the questions you will ask, you must orient those questions toward gaining certain types of crucial information. The areas of this crucial information are the clues to your interview goals, and they must be discerned first. For example, your interviews during feasibility will concern problems and resources, and your interview goals will have to encompass both those information areas.

FIGURE 4.1

Guidelines for Interviewing

A. Before the interview
   1. Arrange for the length of time you need (asking for too much may delay the interview).
   2. Arrange for the interviews to be close to each other in time to minimize cross-talk between interviewees.
   3. Set up a time that will be relatively free of interruptions.
   4. Avoid interviewing two or more people at once (otherwise, there may be too much extraneous data and domination by one interviewee).
   5. Ask in advance if it is permissible to take notes.

B. During the interview
   1. Be punctual: don't be late and don't overstay your time.
   2. Be ready to construct on-the-spot questions that clarify the interviewee's remarks.
   3. Listen to the tone in the interviewee's voice.
   4. Observe body language and facial expression.
   5. Don't wander off the subject; get to the major points.
   6. Don't bias the results of the interview with leading questions.
   7. Don't rush forward with solutions.
   8. Don't easily accept solutions put forward by the interviewee.

C. After the interview
   1. Write up results of the interview as soon after the interview as possible.
   2. Send to the interviewee a note with results of the interview to avoid future misunderstandings; if possible, read the results to him or her before sending the note.
   3. Call after the interview has been conducted; thank the interviewee for his or her time.

You should approach your interviews with the following things in hand:

1. All the background information you'll need.

2. An outline of desired information (interview goals).

3. An open mind.

4. A tape recorder or note pad.

5. A good memory.

Interviews should be scheduled well in advance to allow the subject to refamiliarize himself with the information he'll have to impart and to ensure that he'll have enough time and concentration to devote to the interview.

While conducting an interview (which should be done in person, by the way, and not over the telephone), follow the interview rules in figure 4.1 religiously. Use item 5 of the interview tools just listed to record body language, reticence to commit to answers, approval or disapproval of questions, and cooperation.

Above all, use item 3, your open mind, to allow the interview to do its work. Don't use an interview to validate a personal theory. Doing so will warp your questioning and limit the flow of information. It will seriously weaken the fetal system you are trying to guide to maturity.

Colbert Gathers Information

Let's suppose that such a process is carried out in the Versailles case study. Colbert has accepted his commission to solve Louis's political and economic problems and is embarking on a fact-finding mission to determine a probable solution. His first step is to accumulate (personally or by means of a research staff) whatever pertinent information he can find in libraries, archives, tax records, heraldic colleges, and previously compiled reports. Armed with a solid background of information from these traditional sources, he lays out a program of interviews leading up to another interview with the king.

Using the division of Louis's problem into political and economic components, Colbert schedules the types of interviews listed in figure 4.2. Note that each interview has separate and distinct interview goals; each includes specific questions that will lead open-minded Colbert to a conclusion at the close of this chapter.

FIGURE 4.2

Colbert's Interview Objectives

POLITICAL ARENA

1. Interviews with powerful nobility
   a. What is the source and extent of your income?
   b. What is the basis of your hereditary claim to privilege?
   c. What is your rank in the kingdom?
   d. Who are your allies?
   e. Is glory or power more important to you?

2. Interviews with judicial officials
   a. Which appeals do you accept?
   b. Which courts have jurisdiction over you?
   c. What kinds of cases do you try?
   d. When was your court founded, and by whom?
   e. What is your territorial jurisdiction?
   f. How are justices appointed?

3. Interviews with government bureaucrats
   a. Who is your superior?
   b. What is your social status?
   c. What are your duties and jurisdictions?
   d. To whom is your allegiance: the Church, the nobility, or the king?
   e. What is your tax status?

ECONOMIC ARENA

1. Interviews with tradesmen
   a. What is your income?
   b. What is your tax status?
   c. How stable is the marketplace?
   d. What incentives do you need?
   e. Do you sell French or imported goods?
   f. To whom is your allegiance?

2. Interviews with artisans, craftsmen, and artists
   a. What commissions are you working on?
   b. What skills do you have?
   c. What skills can you develop with incentives?
   d. What incentives do you need?
   e. What is your income?
   f. What is your tax status?

3. Interviews with importers
   a. What is your income?
   b. What is your tax status?
   c. What kinds of goods do you import?
   d. From where do most of your imports come?
   e. Who are your customers?
   f. How much capital have you invested abroad?
   g. To whom is your allegiance?

Each interview becomes background information for the one that succeeds it. Consequently, the chronological sequence of the interviews takes on unexpected importance. Colbert couldn't possibly ask intelligent questions of the wholesale merchants dealing in carpets and tapestries if he hadn't already interviewed the carpet makers and tapestry weavers. Likewise, you can't possibly ask properly pertinent questions of data processing supervisors, for instance, if you haven't already interviewed their customers for information on the quality and promptness of service provided. This sequencing of interviews is an important part of your interview game plan. Proper attention to scheduling interviews helps avoid:

1. Interviews that are invalidated later through new information.

2. Repeat interviews covering the same ground.

3. Interviews that are incomplete in scope or depth of information.

The last step in the interview process is to verify the information you've obtained. You do that in two ways:

1. Check it against other information.

2. Check your notes and memories with those of the interview subjects. This can be done by memo in most cases: "My understanding of last Thursday's discussion is as follows. . . . Please get back to me if you are not in agreement."

The Old System: Analysis and Good-bye

Hold on. We're about to cut our ties with the past—something never to be done in haste or without proper consideration. And we need an accurate record of the past to refer to in the future. After we document the old system, we'll begin the long construction of the new one, even though we may have to let the old system continue to chug and choke away during the time we consume in developing a solution.

The final documentation of the old system is a summary of the information we've gathered, together with the information supplied to us in the interviews. It should contain the following things:

1. A usable schematic of the old system.

2. A summary report of information collected, together with notice of where such information is filed, should it be needed later.

Figure 4.3 is a simple schematic diagram of a two-component system. Every industry, every art, and every discipline has its own method of constructing such a schematic depiction of what's been going on. There's no specific need to make a graphic schematic, if that's not your cup of tea. Your schematic description can be in Urdu prose, if that's the common jargon of your profession. Many systems analysts find graphic schematics useful reference tools—and they do lend themselves more gracefully to wall mounting than do prose reports. But don't feel inadequate if you can't construct or read complex flowcharts; you aren't alone. And there's no reason why flowcharts must be a component of system development—if you have an alternative that works better for you and your group. Despite that disclaimer, note the following things about figure 4.3:

1. It details all flows of information.

2. It details all possible recycling of information.

3. It depicts a continuing (rather than one-shot) system.

In developing your schematic depiction of the old system, remember that a system is a group of interrelated components. Each component receives input from somewhere, processes it, and coughs up some form of output, which will be passed on to the next component. Your schematic should detail each input, process, and output segment of the system.

Devising Alternatives: Woolgathering and Angels' Voices

Well, there you sit. You know everything you need to know about what has been happening in the past. You know everything you need to know about the user's symptoms and resources. And suddenly, it appears frighteningly clear that you know nothing at all, because it's at this point that you do an about-face. You rotate from backward-looking to forward-looking—and there's very little ahead except unexplored space.

How is creativity sparked and carried out? People have been trying to answer that question for centuries, and far be it from us to try to second-guess

FIGURE 4.3

General System Diagram

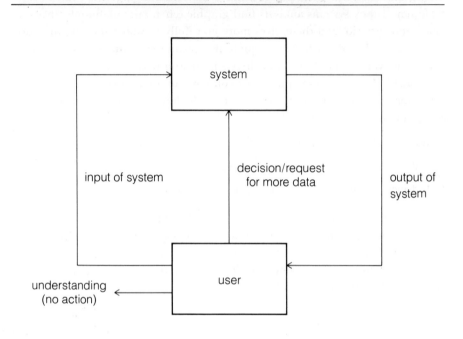

Let's look at the figure to see what it tells us. There are two items in boxes: system and user. They are components. Each component is composed of an input, a process step, and an output. A system is a set of interrelated components. The system in the box is a subsystem of the larger system also comprised of the user. The boxes are *components*. The arrows indicate the flow of information between the components. One flow is the *input* from the user to the system. Another is the flow from the system to the user—the *output*. There are some other arrows, which emanate from the user box. Once the user gets output from the system, something happens. In one case the user understands the data and takes no action. In another case the information alerts the user to possible danger; the user then asks for additional data. In the last case the data causes the user to make a decision that provides a control to the system.

So there we have it. How did we come up with this system? Get in a car sometime. The system in the box is the car. You are the user. If the oil warning light comes on, you are on alert and will pull the car off the road to inspect the oil reserve. If no light comes on, you drive on.

da Vinci, Locke, Kierkegaard, et al. There's no way to describe the ways in which the human mind creates ideas.

That fact is little consolation while you stare at the data and try to devise alternatives. But here are some hints to help you along the road:

1. Write everything down.

2. Work out the obvious and the mundane alternatives first.

3. Don't be afraid to relax and daydream—as long as your daydreams stay in the subject area.

4. Work out each alternative in enough detail to be able to evaluate it later.

5. Check each alternative against the group I list to see how efficiently it will solve groups of problems.

6. Check each alternative against the tentative group II list to see how well it jibes with the objectives you agreed on with the user.

7. Don't be afraid to write down ideas that may seem silly or impractical to you. Remember that you won't have to show your notes to anyone, and sometimes truly fanciful ideas will lead your mind to something workable when followed through.

That's a commonsense list of helpful hints, and it probably could be extended to several pages. It should prove helpful if followed closely. Trust yourself to solve the problem—in conference with associates if necessary or practical. And be sure to cover the whole range of problems listed in the group I and tentative group II lists. Attention to the group I list will focus attention on hard-to-solve aspects of the user's problem and will make obvious the aspects of the problem that are easily (or relatively easily) solved.

The alternatives you develop are based on the group I list, group II list, and the environment. In some fields the flexibility can be limited by cultural constraints. The amount of free choice in these situations can be very limited. Alternatives can be shaped by tradition, culture, and sentiment.

Colbert's Alternatives

Colbert, now appointed to a post equivalent to a modern commissioner of public works, sits scowling in his study at the Louvre and works out a list of alternative solutions by candlelight. Figure 4.4 presents the list that Colbert devises after having cogitated long and hard on Louis's group I list (see figure 3.3, page 51).

FIGURE 4.4

Colbert's List of Alternatives for Louis XIV

ITEM	DESCRIPTION	IMPRACTICAL? WHY?	DEALS WITH WHICH GROUP I ITEMS[1]
1.	Start a war with Spain	Outcome uncertain; cost uncertain	1, 3, 5
2.	Move government to royal palace at Fontainebleau	Palace outdated; nobility has area staked out with own land holdings	———
3.	Establish new ''supreme court''	Possible	2, partly 5
4.	Solidify power through intrigue and politics	Tried before, marginally successful	1, 5
5.	Use position to displace power of nobility	Very risky; remember the Fronde	1, 3, 5; negative impact on 6
6.	Move government out of Paris to new building site; invite nobility to live there on Louis's terms and with financial incentives	Cost uncertain	1, 3, 5, 6, 7, 8, and 4, if 7 and 8
7.	Use only French products to build with	Cost uncertain; some goods not available	4, 6
8.	Use Church to displace troublesome nobility	Tried before, backfired: helpful nobility became more powerful than ever	3
9.	Find new sources of income by colonial expansion	Outcome uncertain, risk of war, cost uncertain	4, 6, 7, 8
10.	Impoverish nobility by taxation	Would start revolution	4
11.	Appeal for help to foreign monarch	No ally is trustworthy except Poland and Russia, which are too remote; risk of war or invasion	1, 5

1. These numbers refer to the group I list, page 53.

Such a format for working out alternatives may be helpful in many situations. It's especially helpful to keep a column on your work sheet to note which group I environmental pairs are being dealt with by each proposed action.

Note that some of Colbert's alternatives are, in the terms of the day, patently absurd. The tenth alternative on the list ("Impoverish nobility by taxation") would no doubt accomplish the overthrow of the monarchy in record time. The nobility always had been—and would continue to be until the Revolution—totally exempt from taxation.[1] Nevertheless, this alternative plays its part in Louis's final solution. Although the impoverishment of the nobility is not accomplished by taxation, it is finally accomplished by encouraging luxurious living, gambling, and dependence on the royal dole. Noting such an alternative is not without reason if its potential impact is great enough.

To some extent, the most absurd of all these alternatives also played an important part in the remaking of the French state under Louis XIV. It appears on Colbert's list as number 11. Louis's government even formed an alliance with Cromwell's antimonarchist England to enhance the power and prestige of centralist government in France. And nothing would have been more dangerous in a less planned environment than for Louis to give tacit treaty approval to the men who killed his cousin, Charles I.

Evaluating Alternatives

Once you develop a list of alternatives and work them out in enough detail to evaluate them, you must undertake a comparison of them. That comparison will produce a preferred alternative—the one that will shape the system solution. Needless to say, the process of comparison, of evaluating the alternatives, is yet another crucial step in the development of a successful system.

Briefly, there are five steps in the process of evaluation:

1. You must develop criteria—yardsticks by which to measure the alternatives you have developed.

---

1. Except in three provinces: Dauphiné, Provence, and Languedoc, but these represent minor exceptions.

2. Develop a method for measuring the criteria against the alternatives.

3. Apply the criteria to the alternatives analytically.

4. Select a preferred alternative or combination of alternatives.

5. Report your findings to the user for approval.

Let's take those steps one by one, because each of them is important to the sequence of events. They represent a chain effect: the quality of each depends totally upon the quality of all those preceding it. If one step is sloppily or inadequately performed, it becomes the weak link, which will snap apart with a big bang later on. If that snap occurs, be assured that there will be retribution; in Louis's time, one could have said that heads would literally roll.

Developing Criteria

It would be nice to lay out a list of criteria by which to measure all alternatives. Unfortunately, it isn't that simple, because every system differs from every other system, and every problem has its own special facets requiring particular yardstick measurements. It would be folly for us to lay out criteria that would accommodate systems as widely diverse as academic reorganizations, missile fuels, and market research.

Nevertheless, there are areas that you should investigate first. The areas that usually demand first consideration in the development of measuring criteria are:

Cost.[2]
Time/scheduling.
Reliability.
Security.
Quality.
Error rate.
Volume.
Profitability.
Flexibility/adaptiveness.
Mobility.

---

2. Note that *cost* is stressed; it is frequently an overriding criterion.

Savings.
Revenue.
Consumption of resources.
Beauty.
Environmental impact.
Public acceptance.
Ease of implementation.

This list is by no means exhaustive, but it should start a thought process. Obviously, some of the above won't apply to all systems: an administrative reorganization won't concern itself with beauty; a construction project probably will have no interest in the mobility of a new office tower. Don't feel obliged to apply each of these criteria to your projects; on the other hand, there may be other criteria that are of overwhelming importance to your system.

Generally speaking, you can develop a satisfactory list of criteria by analyzing the information supplied to you by the user. After all, it will be his baby when it's finished, so you're acting as his representative in developing yardsticks to measure with. Look back over the group I list in developing criteria; make certain that all the user's needs are covered. If your user is more than normally concerned with cost, you probably should consider breaking cost down into more categories: cash flow, capital requirements, loss of interest, and availability of capital from traditional sources. When you have developed an adequate and accurate list of criteria, you're ready to develop a method for applying it.

Methods of Evaluation

We could easily spend several hundred pages telling you all the accepted methods of evaluating alternatives—of applying criteria to lists of possible alternatives. Since this text is not primarily a text on evaluation, we have chosen two common methods, which are illustrated in figures 4.5 and 4.6. As with the list of possible criteria above, our choice of two methods is somewhat arbitrary. There are many others.[3]

---

3. Books on technical evaluation methods include: Grant Ireson, *Principles of Engineering Economy* (New York: Harcourt, Brace World, 1969); and Gerald A. Fleischer, *Engineering Economy* (Boston: Allyn and Bacon, 1972).

FIGURE 4.5

The Net Present Value Method

Suppose we wish to compare two alternatives (A and B). Each alternative can be viewed as an investment that lasts several years and that may or may not have some scrap value at the end. The net present value method assumes we can quantify the benefits and costs that occur in each time period (usually specified in dollars). Getting these numbers is part of the feasibility stage. We'll now assume that we have done this and have obtained the following table.

	ALTERNATIVE A				ALTERNATIVE B		
PERIOD	CASH INFLOW	CASH OUTFLOW	NET CASH FLOW		CASH INFLOW	CASH OUTFLOW	NET CASH FLOW
0		−100	−100			−100	−100
1	60	−10	50		80	−10	70
2	60	−10	50		90	−10	80
3	70	−10	60		20	−10	10

The columns headed by Net Cash Flow are merely the difference between cash inflow (benefit) and cash outflow (cost). Each of these alternatives returns a net amount of $60 over the three-year period. You prefer B. Why? It returns your money faster. The net present value method bears this out. It takes into account the value of money over time. To consider this we'll need some notation. Let

N = number of time periods
I = interest rate
P = present value today
F = future value after N periods

Now if you deposited P dollars in an account earning at the rate of I, our value at the end of one year (F) would be

$$F = P(1 + I)$$

In general, after N periods of time the future value would be

$$F = P(1 + I)^N$$

These mathematical equations hold providing that cash flows occur at the beginning of each period. To get the present value P we can just perform division to get

$$P = F/(1 + I)^N$$

This is the formula we will use in our example. We apply it to each net cash flow for alternatives A and B in each of the three years. Let's set the interest rate I at 10 percent. The results can be tabulated and are shown below. Some sample calculations are

$$45.45 = 50/(1 + .10)^1 \quad \text{and}$$
$$66.12 = 80/(1 + .10)^2$$

PERIOD	PRESENT VALUE ALTERNATIVE A	ALTERNATIVE B
0	−100	−100
1	45.45	63.64
2	41.32	66.12
3	45.08	7.51
total	31.85	37.27

Note that we do indeed prefer B over A.

We've termed the two methods "net present value method" and "scoring method." The former is commonly used in business and other cost-conscious industries. The latter is commonly used in government and in situations where imponderables weigh heavily in the selection of a preferred alternative. There are drawbacks to both methods—as to all methods. The best evaluation methods add several pounds of intuition and horse sense to a few ounces of formal method. But each of these methods does have a structure to work within, and either one can help you accomplish a satisfactory evaluation of a list of alternatives.

The principal drawback to the net present value method is that it measures accurately only when used with firm units such as dollars. It's a sound method when applied conscientiously to investment systems, long-range business plans, budgets of all kinds, financial systems, and data processing systems. It is of virtually no use when applied to situations where important criteria are vague or abstract: beauty, acceptance, artfulness, image, timeliness, and elegance.

The scoring method is well adapted to situations where abstract criteria are important or to situations where a single criterion not measurable in small units is predominant. As figure 4.6 points out, the process of "going with the lowest bid" follows the scoring method even though it is primarily a financial criterion.

Whatever method you select, take extra care in setting it up. What you do now will dictate what comes next.

FIGURE 4.6

Scoring Method

In the scoring method we can work with dissimilar criteria. This method is frequently used by government agencies in awarding contracts. A simplified procurement system is shown below. This is how it works. The user organization develops the criteria and includes it in the request for proposal.

Let's see how it works with a simple example. Let's suppose you are the evaluation team. You are shopping for a new car and have narrowed the choice to a Mercedes diesel, a VW Rabbit, and a Ford Granada. Your criteria are initial cost, mileage, repair record, beauty, and road handling. Also included are relative weights between each criteria. Your judgment tells you that the weights for the car are:

Beauty: 2	Repair records: 3
Road handling: 1	Initial cost: 4
Mileage: 3	

Someone else's weights will probably be different. But the car will be yours. The evaluation team then scores each proposal against each criteria. Each score is multiplied by the appropriate weight. These are then added to get a *total weighted score*.

The scores and weights are summarized as:

CRITERIA	MERCEDES DIESEL	VW RABBIT	FORD GRANADA	WEIGHT
Beauty	5	1	3	2
Road handling	3	4	2	1
Mileage	3	5	3	3
Repair records	3	3	3	3
Initial cost	1	4	5	4

In this table, the higher the score, the better. We see that for repair records all three cars scored the same (3). Therefore, we can delete this criteria since it will not affect the ranking. We also observe that the scores of any one car do not dominate the scores of another car. We cannot eliminate any car. We're now ready to compute total weighted scores. They are:

Mercedes diesel: $5 \times 2 + 3 \times 1 + 3 \times 3 + 1 \times 4 = 26$
VW Rabbit: $1 \times 2 + 4 \times 1 + 5 \times 3 + 4 \times 4 = 37$
Ford Granada: $3 \times 2 + 2 \times 1 + 3 \times 3 + 5 \times 4 = 37$

There's a tie. We now resolve the tie by adding new criteria or reevaluation. This method can be modified to test sensitivities by modifying weights—to see under what conditions a Mercedes is favored, and so on.

Suppose we had only one criteria: cost. We would give first prize to the car that had the lowest cost. There are other criteria. But this illustrates the method. It is widely used by such diverse groups as government agencies, consumer groups, and purchasing departments.

A Simplified Procurement System

Organization thinks of problem it wants solved.

↓

Organization develops a request for proposal.

↓

Request for proposal is mailed to potential bidders.

↓

Potential bidders decide on whether to bid.

↓

Bidders prepare and submit proposals.

↓

Evaluation team in organization evaluates and scores proposals on a point basis.

↓

Total weighted score is computed for each bidder.

↓

Bidding finalists are selected and notified; nonfinalists are notified.

↓

Finalists prepare more data and make oral presentation.

↓

Evaluation team makes final evaluation and submits decision for management approval.

↓

Management selects and awards contract to the winner.

Applying the Method

This step in evaluating alternatives frequently fails because it is taken mechanistically. The watchword here should be "flexibility." In applying methods of evaluation, flexibility can be ensured in several ways—each of them time-consuming and dependent on judgment rather than on figures. The easiest way to ensure it is to use a wide range of possible scores in grading each criterion. If you use a scale of 1 to 2 (acceptable to not acceptable), you run the risk of not recognizing gradations of either category. And you force yourself to categorize borderline cases into categories in which they don't belong. A scale of 1 to 5 yields more information and, hence, more flexibility—although this flexibility is based on judgment.

When applying your evaluative method, remember that you have yet to delve very deeply into the user's requirements for a final system. You must choose an alternative by the end of this stage—and the user must approve it— or else the game ends here. And you must choose an alternative that will accommodate what you perceive to be the user's widest field of requirements. Don't lock yourself into a narrow course of action. Don't base your decision to choose one alternative over another on the size of the doorknobs, when the really important criterion is the size of the building. There will be time (and budget) later on to determine the narrowing of the approach. Right now, you want to select an alternative that is adaptable to changes in the environment and to changes in user requirements.

Be very careful in weighting your criteria. Massive systems have failed because criteria were carelessly weighted. The most common cause of failure of this sort is weighting entirely on the basis of cost ("going with the lowest bid") or of some other single criterion.

You can further ensure flexibility by applying the method in phases. Eliminate the obvious undesirables first. Grade the remainder of the alternatives on the most important criteria. Select the few alternatives with the highest evaluative scores. Reapply the criteria, perhaps drawing upon more data from these finalists. Then choose the most likely to succeed.

Above all, avoid the temptation to "let the computer do it." The winnowing out process is not mechanical—at least it shouldn't be. You may want to alter some aspects of some alternatives after your first scoring session; you may want to combine aspects of several different alternatives.

Some companies, schools, businesses, and industries will dictate both the method of evaluating and the way in which the method is applied. In that case, you have little structural flexibility. You must turn your attention to the data you input; make certain that it is carefully sorted, carefully weighted, carefully presented.

Selecting a Preferred Alternative

Once you've finished the application of your evaluative method, you should be able to choose an alternative that seems best suited to the task at hand. It may be one of many or one of few. If it isn't a totally satisfactory alternative, you should consider repeating the whole process, beginning with interviews and data collection, all over again. Reexamine the group I list for accuracy and completeness. Then try again. But don't proceed beyond this point without finding an alternative that you trust, that the user will have faith in, and that has a good chance of coping with all the important aspects of the problematic environment.

Moment of Truth: Report to the User

Now we've been through all our evaluative business and we're ready to report back to the fellow who started it all—the user. More than likely, he's been pushing you all through this stage of development. It's a stage that may well appear to the user to be a do-nothing stage. He's been told that you're studying the problem, but he's seen no results yet. He's committed resources—possibly substantial resources—to this study, but he's still waiting for a product in return. He's likely to be a bit impatient. More important, he's likely to be ready to criticize what you've done rather harshly. Remember that he's the one with the problem, and he's been continuing to live with it while you've been running around collecting data and evaluating alternatives.

So be ready to defend yourself. Don't schedule a meeting with the user until you have all the information you need to answer all his reasonable questions. Don't succumb to pressure and recommend an alternative you haven't fully researched. After all, if Colbert recommended building a palace on a piece of land that turned out to be a swamp, then—despite its excellent commuting distance to Paris—he'd be in deep trouble with the gentleman wearing the crown.

Follow the same rules that you set up for the interview. Schedule the meeting well in advance so that the user can be prepared. Observe body language, unspoken comments, and degree of cooperation. And present your findings concisely, accurately, and adequately.

Colbert Recommends His Preferred Alternative

Monsieur Colbert is granted an audience. He has applied his criteria, evaluated the list of alternatives presented in figure 4.4 (page 74), and decided on a preferred alternative which is a combination of items 3, 6, 7, and 10. He might state it thus:

We recommend that Your Majesty remove the seat of the royal government to a new location. Your Majesty's holdings in the village of Versailles present a most acceptable location. We recommend that a new palace be built on that land, using entirely French products and designs from beginning to end, giving new and powerful incentives to French craftsmen and merchants to expand their skills and trade. The new palace should be the showplace of the known world and should act as a showcase for French industry, products, and taste. The nobility should be invited to live at court at your pleasure only. Since you own all the land for miles in every direction, they will have no alternative but to accept.

The nobility should be systematically precluded from all important governmental offices, except those offices more conspicuous for glory than duties. They should be given government salaries to satisfy their needs and wants and to make them dependent upon Your Majesty for continued existence.

A concurrent reorganization of the kingdom's judicial system should be set in motion, led by the establishment of a new high court of appeals in Paris.

(That's a mouthful to put into Colbert's memoirs—and, in reality, Colbert probably didn't make such a recommendation. Louis did. And coming from Louis, it was not a recommendation; it was law. Nevertheless, Louis used Colbert as his master tool in all important projects, and it was Colbert who oversaw the construction of Versailles. And, by the way, Louis's decision was precisely as premeditated and coldhearted as the recommendation we have devised for Colbert. He set out to geld the nobility with a surfeit of luxury, and he succeeded royally.)

Assuming that Louis accepts Colbert's preferred alternative (and the unavoidable evidence of history is that he did adopt such a course), Colbert is now ready to proceed to the next stage: analysis.

Documenting the Feasibility Stage

Continuing the mounting paperwork connected with any system, we must devise an adequate and accurate way of documenting this stage. This documentation should consist of two main things:

1. A feasibility study, which summarizes the data collected and the methods of evaluation used. It also presents the preferred alternative and the reasons for its selection.

2. Necessary revisions and additions to the group I list, which will represent our goals for the next stage.

The format of the feasibility study will be dictated by the custom in the environment in which you work. It may vary from a brief memo (in the case of a small system within a small environment) to a lengthy tome that includes *everything* you know to date.

The expansion of the group I list should be a further explanation of the items originally presented, together with any new environmental pairs you have encountered in your travels. The revised Versailles group I list is presented in figure 4.7. The revisions to the group I list should reflect both your increased knowledge of the situation and whatever feedback you've received from the user.

FIGURE 4.7

Louis XIV's Group I List

1. C: No control by monarch
   F: Powerful nobility
2. C: Bureaucratic chaos (primarily in courts)
   F: Judgeships hereditary
3. C: Hereditary privilege
   F: Tradition
4. C: Low revenue to monarch
   F: Impoverished and obstructionist taxpayers
5. C: Weak central government
   F: Louis's power base fragmented by recent history
6. C: Flaccid economy
   F: Civil disorder
7. C: Lack of innovation in trade
   F: Foreign trade is aggressive
8. C: Low French trade prestige
   F: No incentive to trade; no products to trade
9. C: Tradition is impeding progress
   F: Tradition is embedded in all current procedures
10. C: There are no incentives to industry
    F: The government is not spending public works money
11. C: French products are not fashionable
    F: Foreign goods have been preferred by the monarchy
12. C: The courts are in constant disagreement
    F: There is no clearly defined judicial hierarchy
13. C: The nobility is too powerful
    F: The power is enhanced and magnified by physical location; that is, Paris has traditionally belonged to the nobility

Finished? not yet

We need to set up a springboard for the next stage. That springboard will be in the form of a firm group II list. The group II list should be a statement of

FIGURE 4.8

Colbert's Group II List

1. C: The government is located at Versailles.
   F: Tradition dictates that the monarch is the center of government; the monarch has moved to Versailles.
2. C: The nobility participates in court life at the king's pleasure.
   F: The palace is entirely Louis's; the nobility has no established prerogatives in the new location.
3. C: The government functions independent of the nobility.
   F: The nobility is restricted to positions of honor only.
4. C: The nobility is financially dependent on Louis.
   F: The cost of living at Versailles is ruinous; Louis is the only source of subsidy.
5. C: French industry is functioning at full capacity.
   F: Versailles needs goods, will accept only French goods.
6. C: There is a court of final appeal in Paris.
   F: Establishment is at Louis's pleasure.
7. C: The government is entirely centralized on Louis.
   F: The government is run by commoners, with the nobility excluded.
8. C: Privilege is ordered to the king's benefit.
   F: New traditions are manufactured for new court etiquette.
9. C: High revenue to monarch.
   F: Increase in tax base due to expansion of industry.
10. C: Stable government.
    F: Policies set by Louis's staff only.

what you want the environment to be like when the system is operative. It's constructed the way the group I list was: a list of conditions generates a list of supportive forces. The group II list is made up of a series of environmental pairs that form the sustaining environment. For the remainder of the life cycle, our objective will be to transform the environmental pairs of group I to the environmental pairs of group II.

Figure 4.8 represents a group II list for the Versailles system. Notice that it's based closely on the preferred alternative approved by Louis. Note as well that the forces correspond closely to the forces listed in the original and revised group I lists.

Exercises

1. Define the following terms:
   a. Interview techniques.
   b. Cost/benefit analysis.
   c. Scoring method.
   d. Schematic view of system.
   e. Feasibility study.

In problems 2 through 6, you are to act as a person who will be collecting data during the feasibility stage. Propose data sources that could be used and prepare a tentative list of questions you wish to ask.

2. A small apartment house has just burned to the ground. Three people were injured. There were five people total in the building at the time. You are to do a study to determine if the cause of the fire was arson.
3. The Chariots just won upset victories in their last four football games. You are assigned to find out why they won.
4. Your son has just come home. He has been in a fight or had some sort of an accident. You want to find out why.
5. The processing plant for movie film has been running over budget and behind schedule. It is decided to do a feasibility study to determine what is wrong. You are in charge of carrying out the study.
6. You have been asked to select a new accounts payable accounting system that works on a computer.

In each case cited in problems 7 through 14, the system failed. Discuss what could have been done in the feasibility stage to avert later failure.

7. The Edsel.
8. Ku Klux Klan.
9. Nazi sabotage attempts in the United States during World War II.
10. American involvement in Vietnam.
11. Red China's backyard steel mills.
12. Harry Fogg was asked to contact the accounting department and perform a feasibility study to determine what is wrong with the payroll system. Checks have been late

and errors have been prevalent. Harry thought immediately that the new state deduction formula was the cause of the problems. After all, it could account for everything. He went to the department and contacted the two clerks who handled the state deductions. He devised several new forms, handed the forms to the clerks, returned to his office, and announced the solution. Too bad Harry Fogg was dismissed from the company two weeks later. What could Harry have done to prevent failure? What political problems were created by his actions?
13. Take a situation of current interest in your neighborhood. There are usually many to choose from. Describe the existing information system. Develop criteria and some alternatives for the situation. You may choose from the partial list of such situations given below, or select one of your own.
   a. Air pollution.
   b. Lack of good public transit.
   c. High unemployment.
   d. School financing.
   e. High property taxes.
   f. Traffic bottlenecks at an intersection.
14. For the problem you have selected in exercise number 13, propose a method for evaluating the alternatives. Discuss application of the method and the data that is needed to do the evaluation.

In problems 15 and 16, discuss whether the described activity is part of the initiation stage or of the feasibility stage. Give reasons for your answer.

15. After a week-long study, it is decided to abolish parking on Main Street between 4 P.M. and 6 P.M. Monday through Friday. Consideration was given only to the time when parking should be prohibited.
16. A month-long study of the marketing department reveals that they have many problems: bad morale, high turnover, low sales, and bad attendance.

In problems 17 through 19, add to the group I list
some appropriate conditions and forces.

17.

GROUP I LIST	**TENTATIVE GROUP II LIST**
C: Dam is inadequate	Build new dam
F: Overflow of water	

18.

GROUP I LIST	**TENTATIVE GROUP II LIST**
C: Lack of information flow up the organization ladder	Redo organization
F: Poor reporting system	

19.

GROUP I LIST	**TENTATIVE GROUP II LIST**
C: Low productivity	Buy new machinery
F: Low availability of machinery	

In problems 20 through 23, develop a list of
possible criteria that could be applied. Discuss
relative weights of the criteria.

20. Selection of a blind date.
21. Selection of a new house or apartment.
22. Selection of a stock for investment.
23. Selection of a method to reduce pollution
from automobiles.

NOUVEAU PLAN des VILLE, CHATEAU et JARDINS de VERSAILLES

Dessiné sur les lieux en 1714, avec la marche que le Roy a ordonné pour faire voir le Jardin, les Bosquets et les fontaines du dit Chateau Royal de Versailles.

5

The Analysis Stage

We have a project. We know where we are now, and we know where we want to end up; our next step is to figure out how to get there. During the analysis and design stages, we'll prepare a road map to get us to a functioning group II environment. The first draft of that road map is the project plan.

Any system, regardless of size, must pause at this point for some important planning. Beginning at this stage, and continuing through the remainder of the life cycle, our costs will escalate dramatically with every step. To ensure that we don't waste time and/or money in this period of rising expenditures, we must have a detailed plan of action which includes our scheduling forecasts, our budgetary requirements, our methods, and our resource consumption predictions.

If that sounds a bit formal, look at it this way. If you're preparing to build a new house, you'll need:

1. Scheduling forecasts, including when each new contractor will be needed, when weather conditions will hamper construction, and when you will be able to move in.

2. Budgetary requirements, showing amounts, dates, and specific allocations of money required to complete the project.

3. Methods, including blueprints, construction materials, and names of contractors.

4. Resource consumption predictions, estimating cash flow crunches, materials schedules, credit maintenance, maintenance and abandonment of your current dwelling.

All these critical elements in the building of your new house must be known in advance of breaking ground.

The first step toward developing such an all-encompassing plan is to decide on a full system development team from strategic decision makers to tactical decision makers. In the case of your new house, that would be:

1. Management: architect

2. Project leadership: contractors

3. Analysis and development: carpenters, bricklayers, plumbers, et al.

Establishing a Team

Assume for the moment that we're on the Colbert level of management in our system development group.[1] We are the strategic decision-making honchos of the team; we're also the primary liaison with our user function. Our first steps in the analysis stage must be to select the personnel to fill the project leadership slots and then to select the analysis and development team under the advice of our new project leaders.

This rule of common sense is frequently broken, and breaking it frequently leads to heartache later on. You may have a cousin whose brother-in-law is a plumber, and you may decide too early to use this family connection to install the toilets. Some decisions like that may be inevitable, but woe to you if your cousin's brother-in-law has a poor working relationship with the men he reports to as project leaders. By selecting the plumber in advance of the contractor, you automatically short-circuit the usual chain of command and create extra headaches for everyone involved.

Obviously, the selection of a system development team goes a long way toward shaping the final outcome of the development effort. If you select an architect with a penchant for Palladian windows, you stand a good chance of

---

1. See figure 2.4, page 35, for a brief review of levels of management.

having Palladian windows in your house. If you select a contractor who owns shares in a lumber company, you might be able to guess in advance where you're going to be buying your lumber. Likewise with any system development team. If you select a computer-oriented staff, you'll probably get a computer-oriented system. More important, if you select a lazy, haphazard staff distinguished chiefly by their constant availability, you'll get the kind of system dictated by their work habits. It's generally a good idea to involve the user to some extent in the selection of a system development team—if for no other reason than to make certain he doesn't hate Palladian windows.

The selection of a system development team and its effect on the eventual success or failure of the youthful system can easily be overemphasized. But it can easily be overlooked in the scramble to get going, too. Take care to take care in your selection of people to work with.

Colbert's Team

Recall the group II list that Colbert compiled at the end of the feasibility stage, and remember that Colbert's system solution must address two different arenas: political and economic. Colbert will need project leaders in the following positions:

1. Creators of the physical plant, which means an architect for the actual building of the new château.

2. Social and political planners, influential and knowledgeable people who can help implement the newly structured roles of the nobility.

3. Judicial experts, who will set up the new supreme court.

4. Economists, who will monitor and guide the setting up of the new French industries necessary to revivify the national economy.

5. Controllers and tax collectors, who will secure the financing and act as watchdogs on the budget.

6. Protocol chiefs, who will set up the operation of the new government and palace living conditions.

We know the names of some of these: LeNôtre, the architect; LeVau, the landscape designer; and Mansart, another architect. The names of some of the others are not as clear. Colbert and Louis probably took on some of these functions themselves and used prominent social figures as A&D members. It is clear, for instance, that Louis used his famous mistress, Madame de Montespan, to encourage reckless, spendthrift living in the nobility—which, in turn, created an unheard-of dependence upon royal largesse for sustenance. The elegant and beautiful Madame became the center of a group of highly placed gamblers-at-cards, a group that assisted in the financial ruin of several noble houses.

It's also clear that Colbert worked very closely with the new houses of French manufacturing: the new porcelain industry, the new tapestry industry, the rug makers, the furniture makers, the wood-carvers, and the fine artists. His liaisons with each of these important new tax bases functioned as project leaders in the economic arena.

The beauteous Marquise de Montespan, vivacious as she was evil. She fell from favor after she reputedly tried to poison Louis—and retired to a convent.

Writing a Project Plan

Once you've assembled a team, you're ready to write a project plan. We should emphasize before going any further that a project plan should not be engraved on stone. It must be a flexible, living document, because it will pass through revisions, reevaluations, and adjustments. Nevertheless, it's essential that every system develop a project plan at the beginning of the analysis stage.

Depending on the complexity and size of the system being developed, the project plan should cover the following topics:

1. Budget. Questions to be answered include: How much money is to be spent? Over what time period? In what stages? To buy what?

2. Schedule for completion.

3. Milestones.

4. Activities needed.

5. Testing plan. How will individual parts be tested? How will they be integrated and tested? When will testing occur? Who will test?

6. Documentation plan. How many documents will be produced? What are they? Who will organize and maintain the project file?

7. Personnel plan. Who will be used in the project? Are they available? When are they needed? When should their services be contracted?

8. Contingency plan. What happens if something goes wrong? What if financing is cut down or off? How likely is it that major environmental pairs will change in nature? Which environmental pairs are most likely to change? What will be lost, in time and money and momentum, if changes occur?

9. Implementation plan. How will the system be implemented after it is built? How will the user convert to the new system? What milestones will mark the beginning of practical implementation?

10. Work review plan. Who will review work for quality? How will they do it? With what standards? How often, or at what milestones?

11. Areas of risk and uncertainty. What happens if it fails partially? What factors and environmental pairs are precarious? What alternatives are available if environmental pairs mutate rapidly?

12. Economic justification. What is the comparison of costs and benefits?

Some of these topics are expendable in small-scale systems, or in systems that have certain peculiarities. For instance, a very low-cost system can successfully

Paolo Soleri's magnificent ongoing vision is apparent at the rising city of Arcosanti in the Arizona desert. The city is a living project plan, and these buildings are highly visible milestones in the architect's futuristic scheme.

delete items 1 and 12. Some simple systems may be able to get by without items 3, 8, and 10. Generally speaking, however, it's a good idea to address all these items in every project plan you write, even if it's to note, "not applicable." Such a complete approach ensures that proper consideration is always given to every important topic.

The Importance of the Project Plan

The project plan has two functions: a roadmap function and a yardstick function. It must tell us how we plan to get to the functioning group II environment; it must also tell us how well we are doing along the way.

The setting up of identifiable, measurable milestones is an essential function of the project plan for both the roadmap and the yardstick uses. Very like a series of highway signs that tell you how far you have yet to go, these milestones also tell you how far you've come. Using such milestones, you can calculate your mileage per hour and your gasoline consumption, and determine how well they measure up to what you had originally expected. Milestones enable you to make adjustments to your budget and schedule. If you suddenly find, for instance, that you are farther away from your destination than you thought you would be at a given time, you can phone ahead and cancel a hotel reservation.

The series of milestones must be complete. For a trip, the last milestone is entry into the city of destination. In a project plan, the last milestone is successful implementation of the system.

Figures 5.1 and 5.2 explain two generally accepted methods of charting milestones and activities, PERT and CPM, and show how schedules, contingencies, and the like are measured against them. As with all forecasting methods, they have their deficiencies; these too are discussed within the figures. By the way, there is no reason why either PERT or CPM must be used by any system development team. They are forms of graphic shorthand used with ease by some industries and disciplines. If you are uncomfortable with them, throw them out. But if you throw them out, be certain to fill the gap with whatever method is accepted by your associates. Don't try to get by without a project plan or a milestone chart of some kind; to do so is to court disaster.

If you are given to map pins, wall charts, and the like, a PERT chart makes a relatively efficient measure-at-a-glance device. It also serves to "keep you honest." It is less easy to fudge schedules a little when they continue to stare at you from the chart in cold black and white.

FIGURE 5.1

Overview of PERT (Program Evaluation and Review Technique)

In the last twenty years a number of technical tools have been developed. These include mathematical models, which are used to consider networks of various types (roads and human systems, for example) and the computer, which has become a widely used and accessible tool. At the same time, projects have become very complex. The building of the first nuclear submarine is an example. Involved were literally hundreds of subcontractors supplying pieces of the final submarine for assembly. A technique used in managing its construction is called PERT (Program Evaluation and Review Technique). The method involves setting up your project as a network. What does this mean? Suppose you want to build a model airplane. Its construction involves opening the box, reading the directions, setting up the parts, building perhaps five small subsystems, assembling these into an airplane, painting the airplane, and, finally, testing it. This can be drawn in a network as shown.

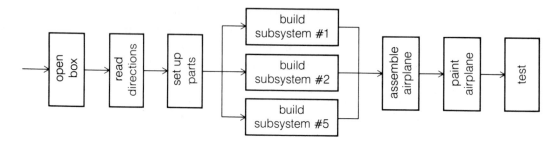

This is one way of viewing quickly the process of constructing the airplane. Each box is an activity.

Another way to view the construction is to look at the boxes as milestones. The arrows and lines refer to activities that produce the respective milestones. Milestones are end products of work (activities). The arrows show the direction of logical flow. An open box precedes the reading of directions. Notice that the charts are sequential in that one thing follows another. However, in the case of building the subsystems, these are shown in parallel. This means the subsystems can be assembled at the same time.

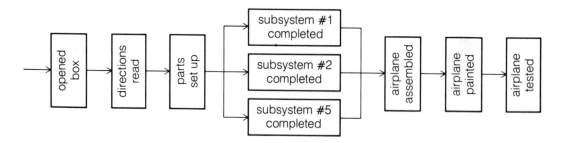

After the network has been laid out, we can begin to construct a schedule for getting to the various milestones. How do we do this? We label the lines in the diagram with the estimated time it will take to get from one milestone to another. This is shown in the next diagram. Obviously, the number will depend upon a variety of factors (complexity, experience, and so on). This is the basic method. It was used on a nuclear submarine project because the building process was so complicated and there was a great demand to finish the project on time. The diagram or network gave project leaders an idea of where the project stood. It allowed them to trade off and place additional resources to speed up part of the project if it was falling behind.

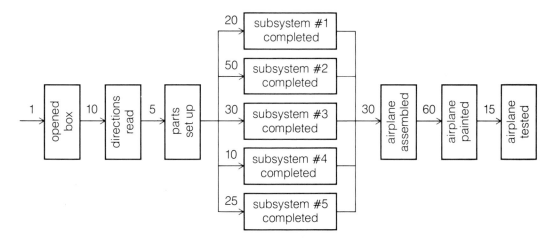

There are some problems with this method:
1. We have to know the milestones and activities very precisely. We need to know which can be done in parallel and which must be done in series.
2. We need to know the times of the various activities.
3. We must know the amount of work required to use the method. You not only have to set up this network, but you must also keep it current and collect data on what actually is going on. You probably wouldn't use it on something as simple as building a model airplane.

The next figure (figure 5.2) discusses how the PERT method can be supplemented by a technique to predict and adjust the completion time of the project.

FIGURE 5.2

Overview of CPM (Critical Path Method)

This method assumes that we have already set up the network and times as in Figure 5.1. The Critical Path Method (abbreviated CPM) is a method to find the longest path from the start of the project to the completion of the project. Let's begin by calculating the longest path in the example in Figure 5.1. All paths from the beginning to the end must traverse the same route except for the five subassemblies. Not counting the five subassemblies, the model will take 2 hours and 1 minute or 121 minutes to build ($121 = 1 + 10 + 5 + 30 + 60 + 15$). The longest path will take this time plus the time for the longest subassembly (50 minutes). The overall time of the longest path is 2 hours and 51 minutes or 171 minutes ($171 = 121 + 50$). Assuming that we can build the subassemblies in parallel, the time to complete the model is a little under three hours (because the other subassembly paths are shorter in time). There are a number of books that describe in mathematical terms how the longest path is found.[1] There are also many computer programs that can develop the network with the data you supply, find the longest path, and present you with graphic results.

In a real project, what do you do when you have found the critical path? Let's look at what happens when something is delayed on the critical path. Suppose that the subassembly takes 55 minutes instead of the originally estimated 50 minutes. The completion of the plane will then be delayed by 5 minutes. In other words, any delay on the critical path delays the project completion.

Let's look at another path. One path takes us through assembling subsystem 3. Subsystem 3 is estimated to take 30 minutes to build. If it takes an extra ten minutes, the completion of the plane is not delayed. But if it takes more than 20 minutes extra, the path through subassembly 3 will be the new longest path or critical path. This provides us with guidance for using the network. We must continually find and reevaluate the longest path to determine when the plane will be completed—and where we should spend additional money and effort to speed up progress.

1. References on network analysis, PERT, and CPM include: B. P. Lientz, *Computer Applications in Operations Analysis* (Englewood Cliffs, N.J.: Prentice-Hall, 1975); Harvey Wagner, *Principles of Operations Research* (Englewood Cliffs, N.J.: Prentice-Hall, 1969); and Frederick S. Hillier and Gerald J. Liebermann, *Introduction to Operations Research* (San Francisco, Calif.: Holden-Day Publishing, 1967).

How to Compile a Project Plan

The responsibility for the completion of a project plan falls squarely into the lap of the strategic decision-making role: management. In a simple system, that presents no problem; either the task can be delegated to the project leader, or it can be accomplished by the management role.

In a complex system such as the Versailles case study, the process is not so easy. Colbert and his staff would face an insurmountable problem in writing a project plan by themselves. What happens in a case like this is that each project leader writes a project plan for his project area. Then the management role assembles the final total project plan from the various pieces submitted by the project leaders.

This job of collating, evaluating, and assembling will be typical of the documentation process throughout the life cycle of a complex system. Strategic decision makers should become familiar with the process; they need to feel at home with it. The development of a project plan, dealing as it does primarily with internal information, fits well into the category of managerial decisions (as does other documentation). Any strategic decision maker who can't delegate such essential tasks will find himself overloaded with paperwork, dissatisfied subordinates, and, eventually, incompetent project leaders (who have abandoned their decision-making prerogatives in the absence of properly delegated responsibility).

Project Plan Review

So there you are. You've collected and compiled a prodigious project plan that provides for all contingencies and which is a product of forecasting and decision making.

Follow the same rules for arranging a project plan review as for scheduling any user interview. Schedule well in advance, prepare any and all information that you may need, and prepare interview goals. And we have a new caution this time: *Talk to the user in his own language.* Remember that this is the first time the user will have to deal with a product put together entirely by your system development group. Jargon, shorthand, and technical lingo have no place in a project plan review. They can only serve to confuse the user. And remember that a confused user will most readily say no; he may also easily make bad decisions in an effort to cope with surface obstructions.

If your system development group deals primarily with computers, and you're dealing with the owner of a shoe store who wants to revise his payroll system, *don't overpower him.* You may know (should know) a lot more about computers than he does, but he knows a lot more about shoe stores than you do. Give him the information straight—and as simply as possible without oversimplification. If he wants to get into the technical background, he'll ask for it. If you confuse him, he won't trust you.

Communication with the user function now becomes a major task for the system development function. The user has hired you because you know more than he does; however, he still retains all the final decision-making options. It's part of your job to make totally certain he understands what's going on. If you lose touch now, heaven help you later on.

With that caution under our belts, let's assume that the project plan is reviewed as written, and that it's approved. In Colbert's case, we can assume that Louis gives a no-holds-barred, all-the-stops-out OK. Louis is that kind of man.

We Forge Ahead: More User Requirements

Once you've won approval of your project plan, you're well on the road to creating a viable system. At this point, you and the user have agreed upon a preferred alternative (the feasibility stage), a method of attaining it, and a set of milestones to tell you how things should happen. You have gathered a system development team together, at least in skeleton, by appointing a project leader or group of project leaders.

Our next focus will be on detailed user requirements, an exact description of the group II environment we compiled at the end of the feasibility stage. The techniques we will use in gathering detailed user requirements will be similar to those we have used before in collecting data but much more exhaustive. Working with the user function, we must now create a total picture of the finished, working system *in its effects*. For the moment, we won't worry about the way the palace at Versailles will be built or what it will look like. We'll concentrate on what it will accomplish—and how it will accomplish it.

Take another look at figure 4.8, page 87. Colbert's group II list gives us a good general picture of the king's requirements for our Versailles solution. It does not, however, give us many of the hows and whys of the project. That's what we are about to assemble. But first, another word about failure.

Another Word About Failure

If we seem to be a bit preoccupied with failure, it's to good purpose, especially at this moment in the life cycle. The user and the system development team are about to part company for a while, and it's essential that we part company in accord.

Often, the compilation of detailed user requirements is approached in a haphazard manner, and the result is presented to the user in a "you know what I mean" kind of presentation. When this happens, there's trouble brewing. The document that details these requirements will be the user's gospel for the remainder of his involvement with the system development team. He'll use his copy of these requirements to test the finished system and to accept or reject it.

That means that in order to protect everybody, these requirements had better be a bit more adequate and a bit more accurate than anything we've previously compiled. An ounce of sloppiness here will turn to an unbearably heavy load later on. It is the responsibility of the system development team to ensure that the user is well-informed about these requirements—to make absolutely certain that these requirements are complete and accurate.

We strongly recommend that these detailed user requirements be written out completely. When the user and the system development team sign off on these requirements, the rest of the life cycle will take its direction and shape from that document. If, later on, anyone needs to refer back to the agreement made, it will be there for quick and easy referral.

Failure will most easily be avoided at this stage by aboveboard communications between the functions. Imagine if Colbert hadn't agreed fully with Louis—in detail—on the proper function of Versailles. Louis might have had to walk long icy halls in search of a mistress; he might even have been installed (heaven forbid!) in a bedroom next to his wife!

In a more serious vein, imagine a payroll system with a sloppy agreement on user requirements. It prints paychecks, transfers funds, and makes no provision for distributing every other Friday. So the paychecks sit there in the payroll department, while one secretary hand-addresses envelopes to 2,500 employees. Not entirely satisfactory, is it? Whose failure is it? It's the failure of the system development team, who failed to find out that no provision for distributing paychecks had ever been made by the user, who was centralizing his payroll function for the first time in his company's history. He thought you were going to take care of it, and you thought—oh, well. It doesn't amount to a hill of beans what you thought, because you've already been fired. And that's failure.

Detailed User Requirements: Colbert

We could easily spend a whole book compiling detailed user requirements on the palace at Versailles. To continue with the bits-and-pieces attack we've used previously, we'll analyze two environmental pairs from figure 4.8 ("Colbert's Group II List," page 87), and demonstrate how user requirements are generated.

Take the second environmental pair, for example. It states pretty clearly that the king's desire is to control the participation of the nobility in the government and that the method to be used to support that control is that new prerogatives will be built into the new palace to restrain the nobility. But with that as a guideline, we'd have a rough time of it trying to design a palace.

So we, as Colbert's system development team, interview Louis and his ministers further. We find out some interesting—and potentially very useful—things:

1. Louis wants the palace to be built to segregate the members of the government from the nobility. He wants to ensure that any duke who wants to hold a government post will have to walk through five miles of corridors to get to his office. Once there, he wants the power-hungry duke to be forced into inelegant, cramped surroundings that seem to imply denial of his exalted rank.

2. The king wants to make the pleasures of Versailles so enticing that no noble in his right mind will want to work.

3. Louis wants to build protocol into the palace physically. He wants to make certain that anyone who wants to see the king will have to pass through certain other chambers. He wants to ensure that dining facilities are such that lower nobility has to wait to eat until after higher nobility has finished. He wants the royal family to have the biggest rooms with the best views. He wants the best parts of the palace centered around himself, and he wants the nobility to have to ask for admittance.

That begins to answer some of the questions. Louis's plan begins to take shape. Now consider the fourth environmental pair. Upon questioning the king, we can elaborate further on it:

1. Louis wants the palace decorated so elegantly that the nobility will have to dress to match it. In the days when silk came from China and pearls were only found by accident, that in itself could break a countess who wanted to show off (it could also break her husband).

2. Louis wants to present magnificent court entertainments. For this purpose, incidentally, Louis is instrumental in developing ballet into a serious art form (although in Louis's day it is considerably grander than it will be in the future). He

The golden age of French showmanship at Versailles. The world still imitates Louis's invention of son et lumière *(light shows); Louis's court playwrights included Molière, Racine, and Corneille. Lully set the mode for opera for 100 years.*

wants his friends to present entertainments, too. He wants anybody who doesn't have a ballet troupe to be considered so coarse that no one will speak to him.

3. Louis is willing to subsidize those nobles who begin to go bankrupt, to supply pearls to the impoverished but gorgeous countess, and to supply table stakes to the duke who takes out his power struggles at the roulette wheel. He's willing to pay for all the ruinous grandeur—if it stabilizes society.

Now, put those two explanations together, and you begin to get a real feel for what Louis is accomplishing by building Versailles. The method is further interrogation of the king (user) and his employees; the product is detail.

User Requirements in the Real World

The approach is much the same, though without some of the glamour and usually without the princely budget. There are two camps among practicing system development groups: those who take the user's word for everything, and those who question the user at every logical point. You must choose for yourself where to place yourself on the spectrum between those two extremes.

You can, in many cases, accept the user's statement of requirements at face value. You can simply accept that he wants to buy a new computer to write his payroll checks. He said it, you didn't. In other cases, the user may be a bit upset if you go along with such a suggestion, especially if he finds out that he could have survived in such a limited situation with a considerably less costly piece of equipment.

The choice between these alternatives is important. Some feel that the total responsibility of the system development group is development of a system and that there's no need to step across the boundary onto the user's turf. There are times when that's true. You may find that the user wants the computer for any number of uses in addition to the payroll system; you may even find that he doesn't appreciate your asking. On the other hand, he may be so grateful to you for pointing out that his golf partner from the computer firm may have been a little overzealous in selling him a new computer that he'll award his next three system development projects to you without a bid.

Of course, not many choices of this type are as obvious as this one. You must decide in each situation which path you'll pursue.

At any rate, you must compile a list of user requirements that describe the system well enough to test it when it has been built. The user will expect to see all his curlicues there when you deliver the finished system. He'll expect it to address all his requirements. He may even expect it to address a lot more than the requirements you note at this point, in which case you can turn the user requirements to your own defense. A complete, accurate list of requirements now eliminates the "I thought you said you would . . ." later on.

Now we have a complete list of user requirements. We're ready to start designing a system. Right? Wrong.

System Requirements—Stepsister

Remember the fairy tales about the pretty stepsister who always gets neglected in favor of the galumphy preferred children? Somehow the compilation of system requirements seems to be relegated to the same position. The temptation to neglect it is great. Who wants to waste time now compiling more requirements? Haven't we already done enough paperwork— feasibility study, project plan, user requirements? Can't we get started?

Remember what we said early on about the advantages of the systems approach. The chief advantage is that we don't commit ourselves or our budget to steps we haven't investigated thoroughly. And at this point, we haven't investigated anything thoroughly except the user. What we want to do now is sit down with the user requirements we just compiled and translate them into usable language. We want to translate Louis's requirement that government be segregated from nobles into a system requirement that they be in widely separated wings of the palace.

We need to make certain that the user's requirements can be followed by technical personnel hired for particular functions. We must translate the user requirements into working documents instead of reference documents. After all, the user requirements were written so the user could understand them. We can't now assume that anything directed toward the user will also be adequate for the personnel on the system development team. They usually speak different languages; that's to some extent why the system development team was hired—the user didn't know enough to solve the problem alone.

It may be that you start the design portion of any project with such an analysis of the user requirements. Certainly, an architect designing a house for a young couple has to build almost entirely from user requirements—without an intermediary. If the user requirements are not complete, the young couple may find that the kids' television backs up to the headboard of the bed in the master bedroom. That means mommy and daddy get to listen to horror movies on Sunday mornings. They'll appreciate the architect and his care in design every week.

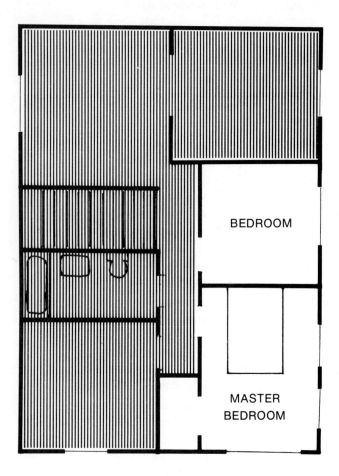

A simplified floor plan of the upper story of a residence. Note that the only available wall space in the master bedroom abuts the children's room next door. A better design utilizing the same space might have provided for both noiseproofing and privacy—for both parents and children.

In a complex system, it's absolutely mandatory that a list of system requirements be compiled before proceeding on to the design stage. And it's essential that the system requirements be checked by the project leaders and the strategic decision makers for practicality, budget conformance, internal compatibility, and so on. (See figure 5.3 for a discussion of a useful technique.) The compilation of system requirements should rank as a milestone in any PERT chart; it represents a turning point in the life cycle.

These four roof styles reflect different sets of user and system requirements: (a) the Mansard-style roof contains a full upper story of the house; (b) the tile roof provides heat insulation and Mediterranean flair; (c) this Alpine chalet retards snow build-up which might crush another roof style; and (d) this flat roof accommodates banks of solar energy collectors.

FIGURE 5.3

Structured Walkthroughs

When you work on a project, you like to have some feedback from your peers. This should be nonthreatening. It won't involve your manager, teacher, or superior, and the feedback can be very helpful. It can make you rethink your position or broaden or narrow the scope of your effort.

A structured walkthrough is an organized method for getting feedback from and giving feedback to your peers. The objective of a structured walkthrough is to find logical errors, stylistic errors, errors that might lead to unreliable or inefficient functioning. The fact that you will be presenting a walkthrough will make you organize and prepare for it carefully. Your audience should be attentive and helpful; after all, it may be their turn next.

What happens in a walkthrough? You present the results of your work. You discuss the logic of the specifications, design, construction, or testing that you did with regard to the system. The meeting is informal. No managers are in attendance. No one is keeping notes or a diary. Frequently no one is in formal charge except the person presenting his work.

How often and when do walkthroughs occur? They should occur very frequently at different stages in the life cycle. Walkthroughs should occur during the development of system specifications, during the system design, during the building and testing of the system, and during the installation of the system.

What are the benefits of walkthroughs? You get help on particular problems. The project group learns what each of the members is doing, the techniques he or she is using, and how he or she is doing. If someone leaves the project, it will be relatively easy to replace him since the people in the walkthrough are familiar with what has been done. Another benefit is that it will save time overall. The alternative is to have each person continue his work without feedback. Fixing errors that are later detected can be very expensive and time-consuming.

What are the problems with walkthroughs? They consume effort. For example, six people meeting in a room for two hours consumes twelve staff hours. Walkthrough results require review to ensure that the walkthroughs are, indeed, finding errors or giving the other benefits cited above. They require considerable initial planning in a group that has previously worked independently.

Where have structured walkthroughs been used? One prominent area has been the development of manual and computerized information systems. A major computer company has cited substantial success in using the technique on several projects. A number of other organizations have had similar experiences.

In general, system requirements may be thought of as translations of user requirements into technical, how-to-do-it terms. It may be useful for some students to think of user requirements as *performance requirements and special features.* In this vein, students may consider system requirements to be *structural characteristics and statements about resources to be used in building.*

If, for instance, the user wants a house built for a family of four, a user requirement might be: the house should be adequate for the needs of four people. The corresponding system requirements might be:

1. The house should have three bedrooms: master, and two children's.

2. The house should have two bathrooms: master and children's.

3. The house should have a den and a family room.

4. The house should have a two-car garage.

5. The house should leave room on the lot for a yard.

6. The house should have two eating areas: kitchen and dining room.

Remember, from here on the documents generated by the system development team will be directed toward A&D members for the most part (although certain reports to the user will also be in order). Consequently, most verbiage from here on will be in the lingo or jargon of the workers, not in the language of the user. The statement of system requirements is both a yardstick and a guide to the A&D members who take over at this point in the life cycle.

Colbert's System Requirements

The compilation of system requirements is the last big step before the project is handed over to the project leaders for design. As such, it represents instructions from the strategic decision-making role (management) to the system development team.

Colbert wouldn't have taken such a responsibility lightly. He was not a trusting or careless man. His system requirements would have been exhaustive, leaving nothing to chance. We discussed earlier Colbert's user requirements for items 2 and 4 on his group II list (page 87). Figure 5.4 gives us an abbreviated version of a set of system requirements created in response to those user requirements. Note that each system requirement is intelligible to people in a particular profession. The architect is addressed in his framework; the economist is addressed in his. Nothing is left to interpretation.

FIGURE 5.4

Some of Colbert's Requirements

USER REQUIREMENT

1. Segregate members of government and nobility. Members of government get less comfortable quarters than nobility.

2. Make Versailles a fun place to enjoy, not to work in.

3. Build protocol into palace.

4. Decorate expensively to require expensive living by inhabitants.

5. Install facilities for court entertainments.

SYSTEM REQUIREMENT

1a. Government and nobility are in separate wings of palace. Those wings are at opposite ends of palace.
1b. Accommodations in government wing are livable but cramped. They don't connect to the king's quarters; officials must walk through the courtyard to gain access to the king's palace.

2a. Build theaters into design.
2b. Expand gardens to include lovers' grottoes.
2c. Expand hunting park to keep nobles on horseback and out of trouble.
2d. Install zoo and exotic botanical gardens.
2e. Install Grand Canal in gardens to facilitate outdoor entertainments.

3a. Arrange floor plan so that layers of nobility radiate out from the king's quarters. Highest nobility is closest after royal family, and so on.
3b. Organize court life around the Hall of Mirrors in the royal quarters.
3c. Order dining areas strictly, with highest nobility eating just after the king in his own dining area. Lower nobility eats in another area.

4a. Use gilt freely.
4b. Use mirrors freely.
4c. Decorate everything; leave no virtue in simplicity.

5a. Build court opera house.
5b. Install spectacular fountains in gardens for night entertainments.
5c. Install facilities for large fireworks displays.

6. Subsidize noble big spenders.	6a. Gather together government monopolies to hand out as large subsidies.
	6b. Channel large portion of tax revenue into king's private purse for personal gifts.
	6c. Revamp heraldic college to check out all claims to nobility and to reduce any subsidy requirements to illegitimate nobles.

Put It All Together: GO or NOGO

Well, here we are at the end of the analysis stage of the life cycle. That means we have run out of allocated budget (because our budget is only allocated in stages) and we have no user authorization to proceed any further. But we are ready to proceed as soon as we get money and permission.

Obviously, we must return to the user (who is also the subsidizer) for the all-important GO/NOGO decision: do we proceed with the design stage of this system project? When we request that decision of the user, we present him with the final products of the analysis stage: the combined user and system requirements (frequently termed *system specifications* in this combined appearance). From this set of specifications, and from your explanation of how it was developed, the user will have to decide whether to proceed, to spend an ever-increasing amount of money, and to invest yet more time.

The interview with the user is critical to both sides; the communication between them must be straightforward and complete. Failure is most easily induced at this stage by the system development team's tendency to insist on an all-or-nothing decision. While this all-or-nothing requirement is certainly not a part of the systems approach, it is frequently a part of human makeup: the system development team sees a preferred alternative with a supporting analysis, and they want to recommend a specific pathway toward solution. That's not how it ought to be done.

During the presentation of the requirements to the user, the system development team should observe the following sequence of information:

1. Review the user requirements for accuracy and completeness.

2. Review the system requirements for their correlation to the user requirements.

3. Review the system requirements for other constraints (such as budgetary and scheduling—the items contained in the project plan).

4. Make any modifications the user requires.

5. Secure approval from the user, along with whatever budget is necessary to complete the next stage (this budget forecast is contained in the project plan).

Don't bludgeon the user. You may find yourself making modifications at this point for a variety of reasons, most commonly because of budget restraints. The user may not have realized just how much it would cost to do all the things he'd like to do; he may have to decide to eliminate some of his more costly requirements to remain within his financial limitations.

Summary

Looking back for a moment, let's review what we have accomplished during the analysis stage:

1. We developed a set of milestones, budgetary constraints, and so on, which we compiled into a project plan. Although the plan will need constant revision, it will be our guide from here on. We secured user approval of the project plan.

2. We collected data, interviewed people, and set down user requirements, obtaining a full description of what the system will do when it's finished.

3. We developed system requirements—technical instructions for A&D members, based on the user requirements.

4. We combined user and system requirements into an approval document frequently called system specifications. We modified it as the user wanted it modified and secured his approval to move on to the next stage: design.

Colbert got the same approval (surprised?). Here we go.

Exercises

1. Define the following terms:
   a. Project plan.
   b. Critical Path Method (CPM).
   c. Program Evaluation and Review Technique (PERT).
   d. User requirements.
   e. System requirements.
   f. Actual versus budget.

In each of the situations in problems 2 through 5, develop a hypothetical project plan outline.

2. Building a house out of wooden blocks.
3. Getting a bill through the legislature from the standpoint of a lobbyist.
4. A day on the job.
5. Painting a room in a house.

The projects named in problems 6 and 7, although technically workable, failed in terms of their project plans. Discuss some of the reasons why.

6. The C-5A cargo airplane (substantially over budget).
7. The tunnel between Britain and France.

In each of the situations in problems 8 through 12, assume that a project plan has been written and approved and that the project is underway. Describe how you would set up a project control system to monitor the costs and schedule of the work.

8. The landscaping of a park site.
9. The building of a new house.
10. The writing of a new computerized payroll system.
11. The development of a five-year economic plan for a local government.
12. The design of a new type of car.

In each of the cases in problems 13 through 16, a project plan is being developed. You are assigned to develop the milestones for the project. You must also define methods for reviewing the work at the end of the milestone to ensure that it is satisfactory.

13. The design and construction of a house.
14. The design and preparation of a manual of procedures for maintaining a car.
15. A trip abroad to visit ten countries in fifteen days.
16. Going to a series of plays.

Discuss how you would develop detailed user and system requirements in each of the cases in problems 17 through 20.

17. You have decided to add a new bedroom and bathroom to your house.
18. You are in the market for a new car. You know from the feasibility stage that you are willing to spend up to $5,000 for a small two-door sedan.
19. You are the mayor of a large metropolitan city and have just received a federal grant of $450,000 to build a new park.
20. You are the police chief of a medium-sized town. A rock festival is being held at a park in the town. You must develop a plan for handling the festival.

In each of problems 21 through 23, a situation is given along with user needs. Develop the system requirements.

21. You want to take a two-week vacation at a cost of no more than $1,200. You want to go somewhere that is sunny and warm. It is December and you live in Saint Louis, Missouri. You want to spend the maximum amount of time at the resort site.
22. You are shopping for a birthday present for an eleven-year-old nephew. He is mechanically inclined. You are willing to spend up to $25 on the present.
23. You are developing a new payroll system. The old one has experienced problems in excessive errors and incomplete input to the system. It has also been late in providing checks.

6

The Design Stage

Now comes the fun. Now we get into a part of the life cycle where we can actually see something happen, something more than just words being put to paper. During the design stage, we'll develop detailed designs for *everything* we plan to do during the building stage. In the Versailles case study, that means we have to design the doorknobs as well as the overall floor plan. We must decide where to get the azalea bushes for the gardens and where to get the bricks for the courtyard, select fabrics for upholstery in the nobles' quarters and wooden benches for the servants, find tapestry makers and tapestry repairers (somebody will have to fix them from time to time after the palace is in use), and locate marble quarries and marble quarriers.

Obviously, the design stage is the beginning of an explosion—in personnel and in budget. From here on we'll be spending money at a considerably faster rate than we have at any point during the life cycle until now.

While our system development headquarters becomes a beehive of activity concerned with design, let's have a brief look at what we have developed so far:

1. We have firm group I and group II lists.

2. We have an approved project plan, telling us (in the user's language) where we ought to be at any particular time.

3. We have an approved set of system specifications.

117

4. We have a skeletal system development team, with project leaders appointed in major areas.

5. We have a budget commitment for the design stage.

Responsibility: Who Does What

The first step in the design stage, during which we'll be dealing with a burgeoning staff and endlessly growing achievement measures, is to allocate responsibility within the system development team. The second step is to ensure constant, complete, accurate communications within the team. The strategic decision-making role (management) must delegate areas of responsibility to each of the project leaders, and must appoint whatever additional project leaders will be required to see things through. As the design stage progresses, staffing requirements may have to be reconsidered more than once.

The A&D role is making its first massive appearance. It did much of the legwork during feasibility and analysis, but decisions were made at higher levels. Now A&D will set to work on the nitty-gritty details of designing the system. Much of A&D's effort from here on will be entirely dependent on each member's individual expertise; much more will depend on the effort made from above to coordinate the separate individual accomplishments.

Coordination must be the key concept from here on. It will become the chief preoccupation of strategic and managerial decision makers. Management's chief responsibility from now on is to ensure that the project leaders work together in harmony, on schedule, and within the budget. The project leaders will have to make certain that their teams interface adequately with other closely related teams and that all work is conducted efficiently and accurately. Gaps in communication during the design stage can result in serious errors and omissions in the final product.

Design: How To Begin

Take the Versailles example. We could sit down now, with our requirements spread out in front of us, and begin sketching floor plans for the palace. We'd have to depend a bit on luck and a lot on memory to get everything into place, and we'd spend a good deal of money on erasers. Nevertheless, some small systems may work just that way: get the requirements together and lay out the design. Figure 6.1 illustrates an interior design problem that might be solved by jumping in and designing.

FIGURE 6.1

Furnishing a Throne Room

Arrange the following pieces of furniture in Louis XIV's Chambre de Préséance at Versailles:

Throne

Dais

Desks (4)

Couches (8)

Flambeaux (8)

Possible solution (one of many):

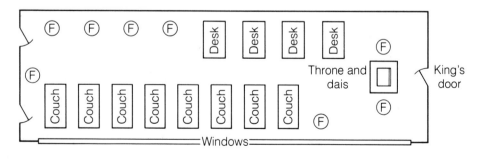

On the other hand, you might want to do a bit of organized brainstorming before you commit your ideas to granite. Sometimes it's best to try organizing things into conceptual units before trying to group them together physically. Remember that tradition, convention, or custom may play important roles in this stage.

Look at figure 6.1. The interior designer has given us a perfectly workable arrangement of furniture in Louis's throne room. The eight couches are arranged to give seating to those awaiting audience with the enthroned monarch. The four desks are available for scribes to take notes on the monarch's decisions. The eight flambeaux are arranged to give everybody sufficient light.

Sounds OK, doesn't it? But Louis will rage at the very idea. That's why, though workable, it isn't logical. Louis's requirements for such a room include:

1. Nobody must sit in the king's presence.

2. The entire room must be ablaze with light.

3. The person seeking audience must feel very alone.

4. The king's presence must overpower the room.

Figure 6.2 gives a design that might be more workable in Louis's final floor plan. It emphasizes Louis's grandeur with the pathway of flambeaux leading

FIGURE 6.2

An Alternative Throne Room Design

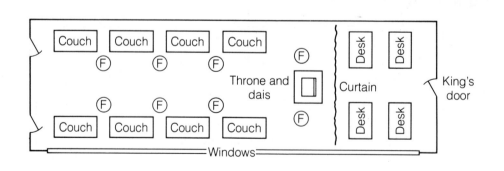

to the throne. The couches are lined up along the walls; they become part of the decor, but are totally nonfunctional—who would stop along the way to sit down? The desks are hidden by a curtain to allow scribes to listen and take notes without being seen. The petitioner has to walk the entire length of the throne room with the king staring at him before he reaches the base of the throne. Much more satisfactory design—for Louis.

How was the design in figure 6.2 worked out? It was not put together until a *logical design* was compiled first.

Logical Design

A logical design is a description in semiphysical terms of what happens (or is supposed to happen) in the completed system. That's a mouthful of definition, but it's really not difficult to understand. A logical design tells you where information moves, where materials move, where people and money move—without really bothering about physical location. It shows the relationships of all the components to each other.

Figures 6.3 and 6.4 are logical designs that might have resulted in the room arrangements given in figures 6.1 and 6.2 respectively. Note that the logical design in figure 6.3 shows the petitioner sitting down after he enters the room, a part of the design Louis would have frowned on. If the logical design had been submitted to the king prior to the finished room design, the king would have vetoed the arrangement before it was ever made.

Logical design has several advantages, the chief one being that it doesn't waste time doing anything more than showing relationships and interfaces. To put that in everyday language, the logical design tells you not where to put things, but where they'll be needed. In the room arrangement in figure 6.2, it's obvious that the couches aren't needed and are used only as parts of the room environment. That fact is reflected in the logical design in figure 6.4. Since the couches aren't needed, they have simply melded into the walls.

Like the techniques we have presented in earlier chapters, the concept of logical design is not tied to a particular form of shorthand. Figures 6.3 and 6.4 use a basic flowchart technique which is easily readable in certain industries. Its premise is that all flows of information (or people or material or whatever) are represented by arrows. There are refinements of this flowchart technique, and there are texts on how to use it. If you're interested, check one out. Meanwhile, decide from the conventions in your industry what method for developing a logical design is best *for you*, and use it.

FIGURE 6.3

Logical Design for Layout of Figure 6.1

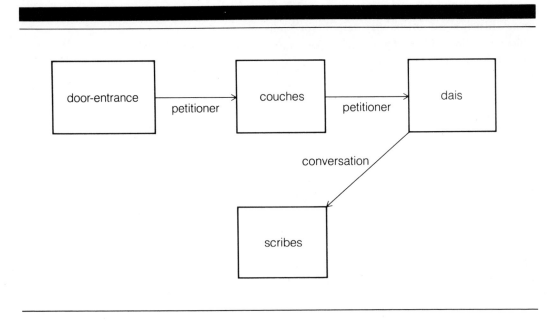

FIGURE 6.4

Logical Design for Layout of Figure 6.2

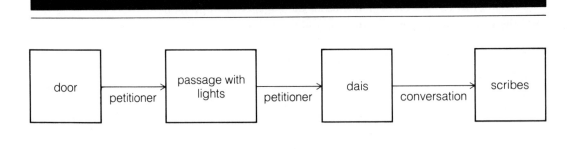

Physical Design—and Failure

With the logical design (in whatever shorthand you're comfortable with) in hand, you are ready to approach the problem of physical design. Obviously, every industry and every project requires different physical design techniques and approaches. We don't propose to include any discussion of physical design techniques here. What we do need to emphasize is the necessary interplay between the logical and physical design processes. Do not think of the logical design as simply an input into the physical design process. The relationship is far more complex than that.

The logical design must act as a check for accuracy and adequacy on the physical design, but there is frequently a back-and-forth process between the two. It may well be that you'll find that physical limitations dictate changes in the logical design, which then affect different aspects of the physical design, which in turn . . . and so on. The relationship between these two design activities is close and critical. Until both are satisfied and in harmony, the project is in a FAIL mode.

Any project which does not have a harmonic logical and physical design can only succeed by accident or by serendipity. Failure results in the design stage from overanxiety, from rushing to get to building, or from an inadequate survey of the requirements. Refer to figures 6.1, 6.2, 6.3, and 6.4. Note that the satisfactory design was derived from a logical design (figure 6.4) translated into a physical design (figure 6.2). The flow of communications and people in figure 6.2 is correct (according to Louis) only as a result of the thought process represented in the logical design in figure 6.4.

Techniques of Design

Wait! Don't just jump in! That's the first inclination, though, and it's one to be resisted. First, you must select a method for your design activities. Choose the approach to the problem of design that will enable you best to meet the design objectives set out in the requirements. And, generally speaking, the bolt-from-the-blue method and the hit-or-miss technique are good ways to court disaster.

Design is a creative process, but, like all creative occupations, it must be accomplished within a framework in order to produce a satisfactory product.

Like a poet who ignores all rules of grammar, composition, and logic, the designer who ignores all the rules will turn out a nonsense design.

There are several angles from which to view a design project: one is from the top, and another is from the bottom. These two viewing angles are designated the "top down" method and the "bottom up" method, and each has its place in the design process.

The "Top Down" Method

The top down method is characterized by concentration on the overall picture rather than on the details. It begins with an overview and concerns itself mainly with integration, harmony, and the overall functioning of the system. Most corporate organization charts are built on a top down view of business life, with all positions hanging from the highest managerial position on the chart.

An example of a top down approach is the common method in new town planning. You lay out the community into large zones, then subdivide it into smaller areas, and so forth. Overall integration is the absolute good in such a case, because the aim of a new town is livability. A transportation system, for example, cannot begin with details; it must begin with a large overview of the area to be served.

The top down method has some particular advantages:

1. Everything stays in its place. Nothing gets blown up out of proportion, because everything is viewed in terms of the total system.

2. System components integrate easier and more naturally, because integration has been a prime consideration throughout the design and building stages.

3. It maintains the hierarchy of importance set for the project. That is to say, the most important aspects of the project automatically receive the greatest emphasis.

But before you run out and opt for a top down approach to the problem, realize that it has some outstanding drawbacks to add to its distinct advantages:

1. Not enough of the details of the system are operational in a top down system until the last piece is in place. It's consequently viewed by many clients as a do-nothing approach or as an unmanageable behemoth.

2. Consequently, it does nothing to solve problems in the short range. If your client is a medical patient who is bleeding to death, the top down approach would be to

begin a total physical examination. By the time you could complete it, your patient might well be dead.

3. Since the emphasis is continually thrown to the overview of the system, the detail work may turn out to be less than satisfactory. Examples are faulty plumbing or malfunctioning air conditioning in new subdivisions, where the only real consideration in design and building was the overall organization of the community.

The "Bottom Up" Method

In its viewpoint, the bottom up method is the reverse of the top down method. The bottom up method advocates immediate solutions, to be integrated later into total systems. In the case of the medical patient we mentioned, the bottom up method would stop the bleeding before making brain scans and doing other processes of general examination. It has obvious advantages in times of crisis.

They're deceptive advantages, sometimes. Short-term solutions are frequently not long-term solutions—as most of us already know. Credit card expenditures may help your cash flow situation today, but not next month when the bill comes in (unless you're expecting a windfall).

The bottom up method has important advantages:

1. It attacks the immediate problem, and begins alleviating symptoms without delay.

2. Each of the components in a bottom up system will work independently of all other components; the detail work will be of high quality.

It has corresponding disadvantages:

1. The system components may or may not integrate when the time comes. The overall picture has not had the greater emphasis.

2. The system may eventually be much more expensive than a top down system because of alterations required to integrate.

3. The system may have entirely the wrong thrust. In a bottom up approach, it is *always* the squeakiest wheel that gets the most oil, and the squeakiest wheel may not be the most critical wheel in the long run.

Structured Design

A close relative of the top down method is a technique termed "structured design." Its workings are explained in figure 6.5. Structured design has found wide acceptance in the worlds of business, administration, and government.

FIGURE 6.5

Some Aspects of Structured Design

Structured design arose out of the need for improved system design in computerized and manual information systems. It was developed by Yourdon and Constantine.[1] It has been used successfully in a variety of projects.

What is structured design? It is a collection of design principles that provide guidance in the design of a system. Let's examine some of these principles. A system can be viewed as a set of interconnected components, pieces, or modules. The art of design is to develop an effective way of structuring each piece and of structuring the relations between the pieces.

Looking at a system, we can detect several principles that will reduce the cost of building and maintaining the system. The pieces need to be kept manageably small. Each piece of the system should correspond to a part of the particular problem. Each relation between two pieces of the system should correspond to a relation between the two parts of the problem. Each piece of the problem and of the system should be solvable and capable of being developed separately. Each piece of the system should be correctable separately and modifiable separately.

These principles can prevent some common problems. Have you ever taken your car in for work and had the mechanic tell you that parts would have to be moved to get at the problem? In one car an engine must be partially removed to get at two spark plugs. In another you must remove the air conditioning to do a tune-up. It hurts in the pocketbook, right? It shows bad physical design.

There are some specific design guidelines that are part of structured design. These parallel management theory. They include:

1. A piece of a system cannot have control over too many other pieces directly. An example follows. The piece labeled *A* has eleven other pieces reporting to it. This can create problems when you have problems in piece A. It may affect eleven other pieces.

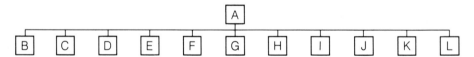

What do we do? We can insert some new pieces (labeled *X, Y,* and *Z* in the following drawing). This will reduce the span of control. These new pieces can be constructed from A or from the other eleven pieces.

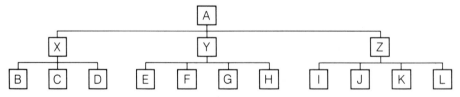

1. Presented in Edward Yourdon and Larry Constantine, *Structured Design* (New York: Yourdon and Associates, 1975).

2. A piece of a system shouldn't have only one other piece in its control. In the case following, either A or B should be dropped and a piece created that combines A and B or puts parts of B into A and C, D, and E.

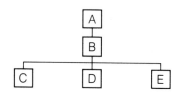

3. A higher-level piece of a system gives only the minimum information or goods to lower-level pieces. The problem potentially here is that giving a piece excess information can create problems. The information may be used in error. It may be changed unknowingly. Have you ever purchased a model or a kit and found extra parts? They created a puzzle, didn't they? Eventually, you probably decided that they were excess. But you might have had that gnawing feeling that you had overlooked something.

4. All pieces affected by the actions of another piece (A) of the system should be controlled by A. This prevents a piece of a system from being affected by another piece out of the line of control.

5. A piece communicates to the pieces it directly controls. It should never communicate directly several levels down in the system. If it does, then the intermediate pieces may not know what is being done at a lower level—yet they are expected to control the lower-level pieces.

6. Horizontal pieces in different parts of the system should have minimum communication. This prevents problems when one of the pieces of the system is modified. Changes made in one module can cause unanticipated problems in other parts of the system.

   These principles and guidelines have implications on how you do a design. Each piece of a system should be highly interrelated and interdependent. There should be minimal interconnections between pieces so that each piece can be treated independently as much as possible.
   Structured design can easily be applied to information systems such as payrolls, inventory control, personnel, and other areas. It also applies to protocol in Louis's court. It applies to a legal reorganization. From our examples of cars, we can see that it applies to physical structures and buildings as well.

The quantitative aspect of structured design makes it easy to get a handle on the areas that need the most work. Its basic principle is that no component should be spread too thin in its responsibilities.

Structured design can be adapted to most systems projects; obviously, however, it is best suited to large-scale projects with multiple components.

Colbert's Design

Like any large-scale building, the château at Versailles must have been designed using a top down method. There is simply no way in which a building could be erected in such harmony and overall grace from a bottom up posture. The architects must have designed an overall floor plan, then narrowed their focus slowly and systematically to include the smallest details.

When one visits the château, one is struck immediately by two things: size and harmony. As a functioning whole, the château works. It works as a piece of architectural design and as a liveable seat of government. Although it is currently in use only as a museum, it has been the seat of several governments, the last having been the Third Republic, which fell at the outbreak of World War II. There's no way to describe adequately the scope of the building. It is monstrous, gigantic, overwhelming. The White House in Washington could fit into it *several hundred times*. Its magnificence is breathtaking even now when tourists are cordoned behind silk ropes and not allowed to walk on the precious carpets.

The protocol design of the château speaks of a structured design technique. No portion of the building is overburdened with people or significance. No petty noble could possibly intrude upon a great noble, because the floor plan is such that it would be impossible for the lesser person to gain access. The château is a labyrinth of chambers, hallways, waiting rooms, presence chambers, and hidden rendezvous. But it works.

Interfaces Between Functions and Roles

At no point in the life cycle is effective communication between the functions more important than during the design stage. The system development function is totally responsible for keeping the user function informed and involved. At each step along the way, the user must be sounded out; his approval must be sought for each important concept. That much is obvious, and, to a certain extent, it's automatic. No one is going to design a throne room for Louis without consulting Louis.

Generally, there are two important checkpoints for user approvals: after the logical design is complete and after the physical design is complete. The user's approval of the logical design is useful to the system development function in two ways:

1. It ensures that the flow of information, personnel, and so on follows the patterns desired by the user.

2. It provides a solid jumping-off place for the physical design. Once a logical design has been approved, the architects at Versailles have in mind particular objectives which jibe with Louis's ideas about protocol, priorities, and so on.

The user's approval of the physical design is required for more traditional purposes:

1. It ensures that the style of building will fit the user's taste, pocketbook, and needs.

2. It ensures that the user perceives the flow approved in the logical design to be well served in the physical design.

3. It gives everyone a final chance to make any changes that seem necessary.

The interfaces between the system development roles are more problematic. Depending on the size of the project, the system development team may be unwieldy and difficult to assemble. Systems of communication and documentation must be set up.

The Problem of Documentation

If there's a danger inherent in the systems approach, it's the danger of creating too many formal procedures. There may be a tendency to drown projects in a flood of required paperwork. That tendency is to be avoided assiduously. But proper documentation is absolutely necessary within the bounds of reason.

You'll recall that the project plan formulated during the analysis stage dealt with documentation in a preliminary mode. But since the full A&D team is only now being assembled, the implementation of a documentation plan is the responsibility of the design stage.

There are two rules that should be observed in designing and requiring documentation:

1. It must be complete.

2. It should not be crippling.

It may be difficult to strike a balance between those two principles, but it must be done. A documentation scheme that ignores the first of the two rules will be woefully inadequate for reference purposes later on; it will also run the risk of leaving segments of the system development team in the dark during important decision-making times. A plan that ignores the second rule will engulf its personnel in paperwork and spend a disproportionately large amount of the user's budget on unnecessary reports.

Many documentation plans ignore both these rules and opt for "doing it the way we did it before." In designing a plan for reporting, keep in mind that every project has different requirements and different milestones. Most projects have different personnel to some extent. Old horses just won't pull the new carriage well.

A proper documentation plan must work closely with the role interface process—a process that is largely informal.

Interfaces

What is a role interface? A dictionary definition might read, "a line regarded as a common boundary between two objects, persons, or functions." Figure 6.6 illustrates some physical interfaces which fit this description.

FIGURE 6.6

Examples of Physical Interfaces

INTERFACE	BETWEEN
1. Wall, doorway	1. Two adjacent rooms
2. Telephone equipment	2. Two persons talking while one thousand miles apart
3. Memorandum, letter, mail system	3. Two people corresponding by letter
4. Jacket, insulation	4. Person and a snowstorm
5. Public relations group	5. A large company and the public

When we speak of a communications interface, however, we're interested not so much in the interface itself as in how to get across it. Figure 6.7 illustrates the major interfaces in the life cycle of a system. Note that only two of the four roles have only one interface: user and A&D. In both cases, these two end stops are where "the buck stops." In the final analysis, the user must swallow all the feedback from the system development team; the A&D members must accomplish the nitty-gritty work of the system.

FIGURE 6.7

Interfaces During the Life Cycle

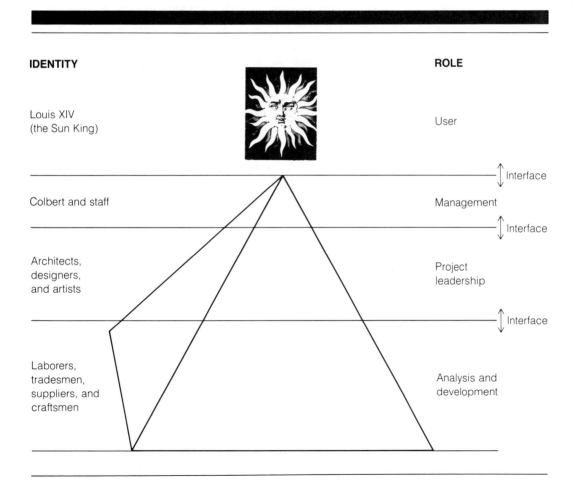

IDENTITY		ROLE
Louis XIV (the Sun King)		User
		↕ Interface
Colbert and staff		Management
		↕ Interface
Architects, designers, and artists		Project leadership
		↕ Interface
Laborers, tradesmen, suppliers, and craftsmen		Analysis and development

Between those two outside roles, however, there are two roles which have more complex interface problems: management and project leadership. Not only do both these roles have two interfaces each (instead of one), but they're primarily responsible for the entire path of communications in the development of a system solution.

The management role is responsible for soliciting information from the user and passing it on to the project leaders. It's also responsible for keeping track of the progress of the system development team via the project leaders and for passing that information along to the user as necessary. Likewise, the project leader conveys information in two directions: it channels directives from management to A&D and moves information from A&D back up the line to management.

Obviously, management and project leadership have more to do than just pass notes back and forth from user to A&D (or they would be hindrances to information, instead of aids). Managerial and strategic decisions are made during the course of a communication from one end of the path to the other.

For example, if Louis (user) tells Colbert (management) to make room at Versailles for Madame de Maintenon (a new mistress):

1. Colbert selects a project leader who is designing chambers for royal mistresses and tells him of the new addition.

2. The project leader (say it's Mansart) fits the new arrival into the logical design and relays the addition back to Colbert for approval.

3. Colbert obtains approval from Louis and relays that OK—with any changes, cautions, or budgetary constraints—back to Mansart.

4. Mansart fits the new mistress's chambers into the physical design, makes a list of any trade-offs that must be made, and relays the list back to Colbert for the same approval sequence.

5. Colbert obtains the second approval and relays it back to Mansart, again with changes.

6. Mansart gathers the workmen who are laying out floor plans in detail; he informs them of the new mistress, where her rooms should be, and how many rooms she should have. He tells them of the secret staircase that should be built leading to the king's bedroom. He tells them her favorite colors and that she has two pet dogs.

7. Mansart checks his segment of the project plan and passes back the design for approval, together with a report on the impact of the new addition on the overall project plan (on budget, scheduling, and so on).

8. Colbert checks the overall project plan and contacts other project leaders who are affected by the change (protocol planners, furniture manufacturers, and others).

9. Colbert secures approval of the design from Louis.

10. Colbert passes the approval with changes to Mansart and to other closely related project leaders.

11. Mansart instructs the A&D team to finalize the design.

12. The A&D team informs Mansart that the design has been finished.

13. Mansart passes that on to Colbert, together with a report on the final impact of the addition.

14. Colbert informs the appropriate project leaders and reports back to Louis.

15. Louis thinks it's great.

Notice how many times the communications interfaces were crossed. Figure 6.8 follows the flow of information, instructions, and so forth. Needless to say, if the communications break down at any point during the addition of Madame de Maintenon's rooms, there might be substantial problems later on.

One of the peculiar problems of communications breakdowns is that they don't come to light immediately. They're usually found much later in the life cycle and are costly because they're discovered during a time of increased expense.

President Johnson (left) had a communications hook-up throughout the governmental network. Despite complex phone systems and teletype machines, communications breakdowns plagued his administration.

FIGURE 6.8

The Flow of Information During Design:
Designing a Room in the Palace of Versailles

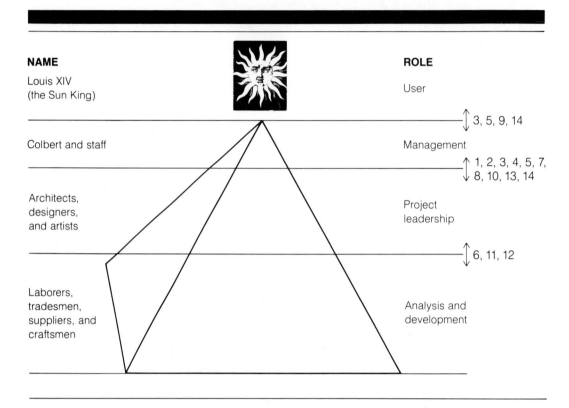

NAME

Louis XIV
(the Sun King)

Colbert and staff

Architects,
designers,
and artists

Laborers,
tradesmen,
suppliers, and
craftsmen

ROLE

User

3, 5, 9, 14

Management

1, 2, 3, 4, 5, 7,
8, 10, 13, 14

Project
leadership

6, 11, 12

Analysis and
development

Relating Back to the Environmental Groups

After the logical design is finished and again after the physical design is
completed, the whole progress must be checked against the group I and group
II lists for conformance to goals and objectives. But before that check can be
made, the lists must be updated. Ideally, the lists should be constantly
updated; they should be checked for timeliness and compatibility at each
designated milestone. Practically speaking, however, it's mandatory that they

be updated prior to the end of the design stage; that's where most budgets will pause for this verification.

First, we must resurvey the user's situation and update the group I list. We must check each environmental pair to see that it still represents the real situation adequately and accurately. Then we must add any environmental pairs that have turned up since our last major survey during the feasibility stage.

Figure 4.7 (page 86) represents our current group I list for the Versailles case study. Colbert must check it over again with Louis for accuracy and completeness. After all, it is the basis of the design we'll be ready to submit at the end of this stage, and that design will have to be modified if any of the group I environmental pairs have mutated.

Some of the pairs may have ameliorated (become less a problem). Take number 11, for example: Louis's ongoing concern for the French luxury industry may have already caused less need for further concern by making French goods more fashionable in the capital. Others may have deteriorated. The process of deterioration may be seen in number 2 of the figure 4.7 group I list; the courts may be in worse, more critical, chaos than ever before.

Each of these changes will affect the final design in some way. The change in number 11 may mean that fewer incentives will have to be offered to French manufacturers (since they are already on the road to recovery). The change in number 2 may mean that a more drastic overhaul of the courts may have to be made than was originally expected.

Also important (and more difficult to attend to) is the fact that new environmental pairs may have popped up during the time we have spent in analysis and design. Louis may have entered into a war with a foreign power (a drain on the economy, but a unifying force for the nation). Still other changes in the group I list may have been dictated by the design itself. The decision to build at Versailles, for instance, may have added to the group I list a new pair, which would be number fourteen:

14. C: The roads to Versailles are totally inadequate to move supplies to the proposed building project.
  F: There has never been any need to move such massive supplies to the hitherto little-known village.

Each of these changes in the group I list must be checked against the group II list for possible impact. New items may have to be added to the sustaining environment to cope with new additions to the problematic environment.

Figures 6.9 and 6.10 represent the updated group I and group II lists as they stand just prior to breaking ground at Versailles.

FIGURE 6.9

Louis XIV's Revised Group I List

1. C: No control by monarch
   F: Powerful nobility

2. C: Bureaucratic chaos (primarily in courts)
   F: Judgeships hereditary

3. C: Hereditary privilege
   F: Tradition

4. C: Low revenue to monarch
   F: Impoverished and obstructionist taxpayers

5. C: Weak central government
   F: Louis's power base fragmented by recent history

6. C: Flaccid economy
   F: Civil disorder

7. C: Lack of innovation in trade
   F: Foreign trade is aggressive

8. C: Low French trade prestige
   F: No incentive to trade; no products to trade

9. C: Tradition is impeding progress
   F: Tradition is embedded in all current procedures

10. C: There are no incentives to industry
    F: The government is not spending public works money

11. C: French products are not fashionable
    F: Foreign goods have been preferred by the monarchy

12. C: The courts are in constant disagreement
    F: There is no clearly defined judicial hierarchy

13. C: The nobility is too powerful
    F: The power is enhanced and magnified by physical location; Paris has traditionally belonged to the nobility

14. C: The roads to Versailles are totally inadequate to move supplies to the proposed building project
    F: There has never been any need to move such massive supplies to the hitherto little-known village

FIGURE 6.10

Colbert's Revised Group II List

1. C: The government is located at Versailles
    F: Tradition dictates that the monarch is the center of government; the monarch has moved to Versailles

2. C: The nobility participates in court life at the king's pleasure
    F: The palace is entirely Louis's; the nobility has no established prerogatives in the new location

3. C: The government functions independent of the nobility
    F: The nobility is restricted to position of honor only

4. C: The nobility is financially dependent on Louis
    F: The cost of living at Versailles is ruinous; Louis is the only source of subsidy

5. C: French industry is functioning at full capacity
    F: Versailles needs goods, will accept only French goods

6. C: There is a court of final appeal in Paris
    F: Establishment at Louis's pleasure

7. C: The government is entirely centralized on Louis
    F: The government is run by commoners, with the nobility excluded

8. C: Privilege is ordered to the king's benefit
    F: New traditions manufactured for new court etiquette

9. C: High revenue to monarch
    F: Increase in tax base due to expansion of industry

10. C: Stable government
    F: Policies set by Louis's staff only

11. C: There are paved roads to Versailles
    F: Road builders are allowed to collect tolls for properly built and maintained roads

Final Update and Approvals

Once the environmental lists have been properly updated and their implications worked into the final design, the project plan must be updated as well. Recall that the project plan is an administrative overview of the proposed system life cycle including budgets, schedules, and so on. The new addition to the group I list (number 14) will necessitate a new budget expenditure for new roads; it will necessitate as well a delay in construction at Versailles until those roads are in condition to move in the granite blocks necessary to build the château.

Similar updates in project plans must be made as a final step in the design stage for all systems. Before approvals can be obtained from users, all impacts of the recommended design must be documented and handed back with the

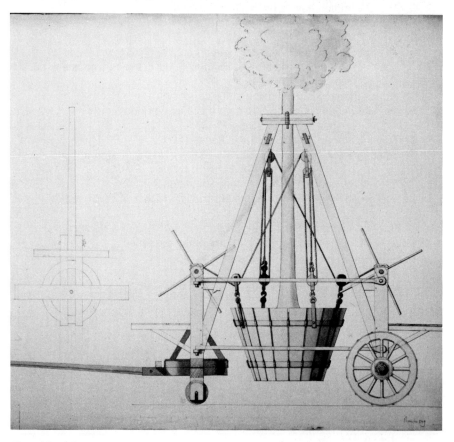

One of Louis's many user requirements was fresh oranges—snow or no snow. Shown is a contraption for moving orange trees indoors, away from harsh Versailles winters.

design. Especially important in project plan updates are budgetary changes, scheduling changes, milestone changes, and scope changes. Without adequate and accurate reporting of changes in the project plan, the system development team would have to proceed on outmoded user authorizations, and, eventually, unexpected expenditures would be handed back to the user, which would anger him. In Colbert's case, that might be a matter of life and death.

When the project plan has been updated, Colbert must return to the series of antechambers that isolate Louis from the troublesome world. He must lay the whole package at Louis's feet and await a reply.

The real world

In the case of Louis and Colbert, the relationship is so close and so heavily dominated by Louis that the approvals are almost automatic. Nothing can proceed far in the wrong direction without Louis reeling it back and putting it on the right track.

In the real world, however, the situation may be quite different. The user may be quite surprised by what turns up for his signature at the end of the design stage. He may have been expecting, for instance, to tie into a time-sharing group for his computer needs. The system development team may have been forced by volume or security requirements to design a wholly captive computer into the final system. The difference in capital outlay may be enormous.

Remember that costs will escalate many times when we enter the building stage and that any modification or turn back will be enormously expensive then. It's best to lay the whole picture out now and get real approvals. You don't help the user by soft-selling the design; his budget won't be able to handle extra costs any better if he doesn't know about them until the bills come in. Scheduling changes are equally important to report now; otherwise the user may be standing holding a melting ice cream cone waiting for you to install the new freezer (figuratively speaking). That doesn't go over well, as you might imagine. As usual, it's the responsibility of management to see that all the updating is done properly and accurately and that the full impact of the design and the updates are conveyed to the user.

Then (hallelujah!) approvals are forthcoming if the system is to proceed. Judging from the reactions of tourists at Versailles each summer, Louis and Colbert were able to come to substantial agreement.

Exercises

1. Define the following terms:
   a. Top down design.
   b. Bottom up design.
   c. Structured design.
   d. Logical design.
   e. Physical design.
   f. Roles in System Design.

2. Look in your local newspaper and find articles on systems or structures that failed. Discuss for each case:
   a. Did it fail due to design?
   b. Who appeared to be responsible for failure?
   c. Could the system have been modified to prevent failure?
   d. Could the cause of failure have been foreseen?

3. You have probably heard of the many recall campaigns that the automobile manufacturers must conduct. Usually a part in a certain model car fails. The cause is often bad design of the particular part. A new one is then designed and built. The recall campaign is conducted and the faulty part is replaced. Discuss what could be done during the design stage in these cases to prevent failure of the parts.

For each of the cases in problems 4 through 8, develop a collection of pieces that comprise the system. For each case discuss how you would approach the problem using (a) top down design, (b) bottom up design, and (c) structured design.

4. A payroll system that computes gross pay and tax, insurance, and benefit deductions.
5. An information system for a candidate for the city council.
6. A three-bedroom, two-bath house to be built on a vacant lot.
7. A survey on voting patterns in an urban area.
8. A field trip for a group of scouts to the zoo 150 miles away.

In problems 9 through 13, determine the information you would need from the system specifications in order to construct the design of the system.

9. Adding a room onto someone's house.

10. Designing a report on items in stock in the hardware department of a large department store.

11. Designing a method for evaluating the use of food stamps.

12. Designing an antithrowaway bottle law in your state.

13. Designing a system for evaluating employees in a county government.

14. You are going to be put in charge of building the systems in problems 9 through 13. Determine what design components you need in order to build each system.

15. For each of problems 9 through 13, assume you are the project leader and are hiring people to work on the design of the system. Discuss the skills and expertise of the staff you would hire to do the design in each case.

16. For each of problems 9 through 13, compile a group I list and a tentative group II list. Explain how you would check your design against the list to see that the designed system would be likely to produce the tentative group II list in each case.

17. For each of problems 9 through 13, assume that several different designs have been developed. Develop criteria and a method for evaluating the alternative designs. Assuming that more than one design satisfies the user and system requirements, how would you select the best system design in each case?

18. A survey is to be conducted in a large state to see if the use of drug XX has an effect on the rate of household accidents. Bill is in charge of the design and reports to Lance. Lance, who has never overseen such a project before, wants the data collection to start in two weeks. It will take one week to design and make copies of any questionnaire. The project team of Bill, Mary, and John feel that they can get a rough draft done in one week.
    a. Discuss why this is likely to fail.
    b. How will failure be manifested?
    c. What can Bill do to prevent failure?

19. A family of five lives in a four-bedroom house. They need an effective but inexpensive way of passing messages from one person to another. There have been too many missed messages without a formal system.
    a. Develop three alternative designs for the system.
    b. Compare each design with the others.

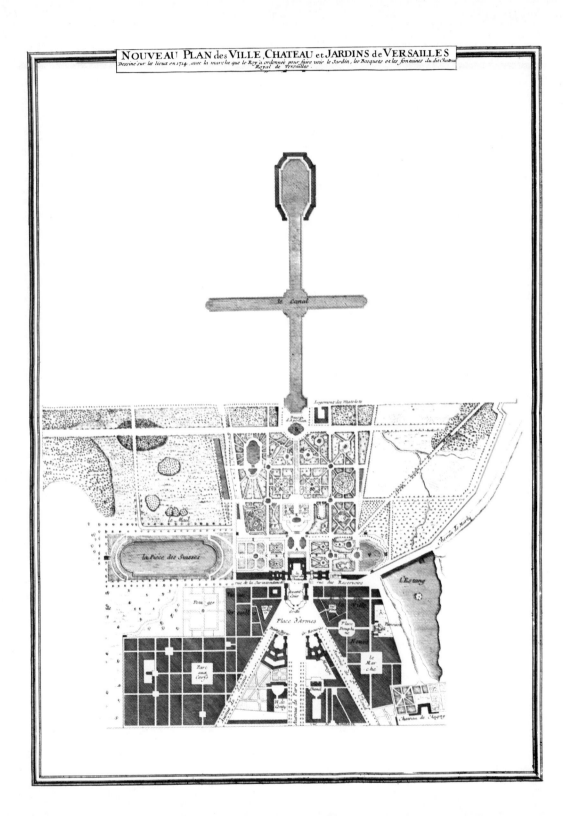

NOUVEAU PLAN des VILLE, CHATEAU et JARDINS de VERSAILLES

Dessine sur les lieux en 1714, avec la marche que le Roy a ordonnée pour faire voir le Jardin, les Bosquets et les fontaines du dit Chateau Royal de Versailles.

7

The Building Stage

Get out your trowels and chisels, your pens and parchment rolls, your needles and thread. We're ready to start building.

Funny, but the common conception of systems is that the work starts at this stage, with some preliminary brainstorming in the way of design. The temptation for system development teams (and certainly for management) is to sit the workers down and set them to work right off the bat. And that temptation, when followed through to a logical conclusion, is the source of a lot of sorry systems.

Imagine, looking back through what we have done so far, what would have happened if Colbert had just started digging at Versailles. Imagine his surprise when he found that he had a caravan of marble blocks that wouldn't roll over the country roads! Imagine his amazement (and the king's) when he established the ministerial offices conveniently off the king's quarters—only to find that the king wanted *only* royalty and high nobility within sniffing range of his area. The backward glance and snicker does not apply solely to the Versailles case study. Imagine the rage with which any management would greet the arrival of a totally unexpected piece of equipment that set them back, say, $250,000. Surprise!

No, the work does not start now; it started before the user ever contacted the system development team. It continued through all our lengthy studies and validating exercises, through our painstaking analysis and structured design. But hold on! It's going to escalate now!

You will see, as you develop a project plan, that there's a nonstop pattern of increasing costs as the life cycle progresses. Figure 7.1–a illustrates the shape such a cumulative pattern will usually take.

FIGURE 7.1

Typical Cost Patterns of the System Life Cycle

a. Typical Cumulative Cost Pattern

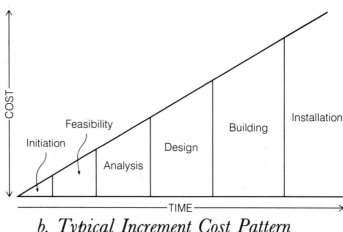

b. Typical Increment Cost Pattern

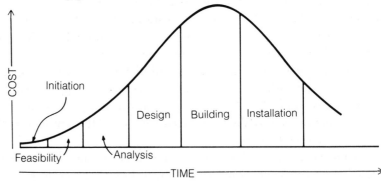

Escalating Costs: Advantage or Disadvantage?

Whether cost escalation is an advantage or a disadvantage depends on how you look at it. The progression of costs can be either an encouragement or a mortality factor, depending on how carefully the system development team follows the logic of the life cycle.

The basic consideration in the development of any system is the environment. We discussed the split between problematic and sustaining environments early in this text, and we've brought the concept into play in virtually every chapter thus far. Now we ask you to look at another side of that dichotomy.

The fact that costs will inevitably rise (and eventually decline in the last stages) as you move through the life cycle is a component of the problematic environment of every life cycle. It is the first item on a universal group I list, and it is the primary concern of every system developer (or should be). This universal group I component is the genesis of the systems approach. In order to convert that problematic component to a sustaining component, the systems approach was born.

Say for the moment that you have never heard of the systems approach, and you want to build Versailles. You attack the problem in the middle, at the design stage (or its equivalent); you hire an architect to start drawing plans. What happens? You have to backtrack constantly as you somewhat chaotically discover the factors that we have organized as user requirements.

For instance, after having poured the foundations of a building, you might discover that the user's study absolutely must have a southern exposure in order for him to carry out his botanical experiments with tropical plants. You never knew that, and he never bothered to tell you, because you didn't ask. What are you to do: pick up the foundation and rotate it ninety degrees? Or you might find, after having built a dam across a wilderness river, that public outcry demands that salmon ladders be installed to preserve the natural environment two hundred miles upstream; no salmon means no bears, no bears means no tourists, no tourists means closing motels, and so on. So you have to delay implementing the dam while you destroy enough of it to install salmon ladders.

There are innumerable examples of nonsystematic approaches and of the potential disasters (especially financial disasters) that they invite. And the reason they invite disaster is that their implementers have not done their homework. Figure 7.2 shows the entry points of the two examples just given on the cost triangle introduced in figure 7.1. Notice that each of them incurs more cost by backtracking than the systems approach would have incurred by starting at the beginning. Both nonsystematic examples have to plod through design and building at least twice, whereas the systems approach would have determined requirements during feasibility and analysis, and would have to pass through the design and building stages only once.

FIGURE 7.2

Entry Points for Rework During the Life Cycle

a. House Foundation

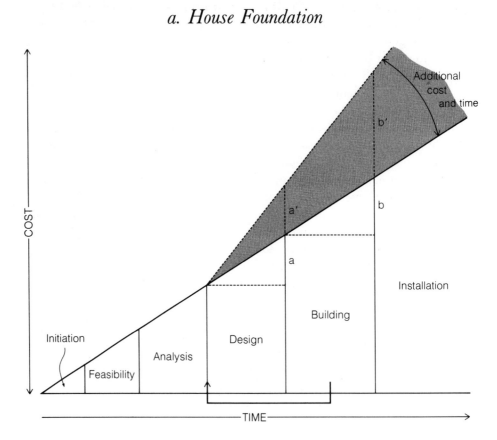

This brief financial analysis of the advantages of the systems approach should also serve to emphasize an inescapably essential fact: *There is little point in following a systems approach if the first few stages aren't very carefully completed.* A shoddy or hurried feasibility stage or initiation stage can effectively put even a system development team in the position of passing through a costly stage more than once. Enough said.

We are now at the beginning of the building stage, which (as figure 7.1 indicates) is a quantum leap in costs from anything that precedes it.

FIGURE 7.2

Entry Points for Rework During the Life Cycle

b. Salmon Ladders and Dam

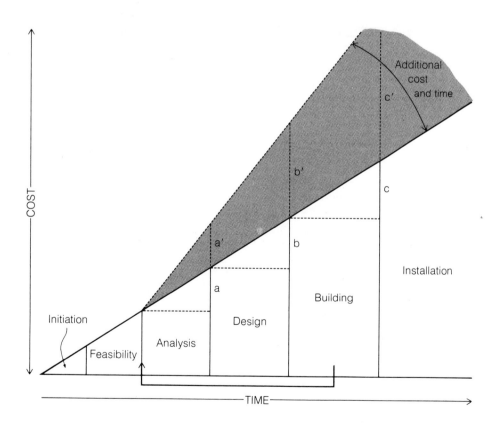

Another Backward Glance

The first task we face in the building stage is to make a detailed expansion of the project plan based on the work completed in the design stage. The expansion of the project plan is our only top down tool in the building stage: it's our only overview instrument to ensure final coordination.

Building can't be accomplished in a top down mode (unlike design). You have to build in a strict bottom up mode, by constructing components and then fitting them together. Consequently, the danger of slippage in the coordinating phase is great. The cartoon in figure 7.3 is an apt comment on what can happen.

FIGURE 7.3

An Example of Coordination Problems During Building:
Laying Railroad Tracks

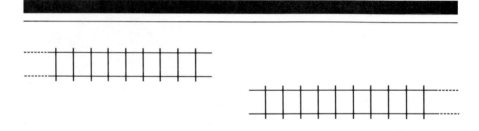

So what do we do? We must analyze the final design in terms of individual milestones and put together a schedule that will tell us when, where, and how the parts must fit together. It might be a good idea at this point to review the discussion of the Critical Path Method, which we presented in chapter 5, page 100. Something along the lines of a Critical Path Method analysis of the activity must be a first step in this expansion of the project plan. You must decide which steps take the longest, which are dependent on completion of others, which must coordinate with others during construction. Figure 7.4 is a drastically oversimplified drawing of what Colbert might expect during the building stage in the Versailles case study. Note that the political and economic aspects are drawn along with the physical construction aspects of the palace.

FIGURE 7.4

Simplified Project Plan—Milestones

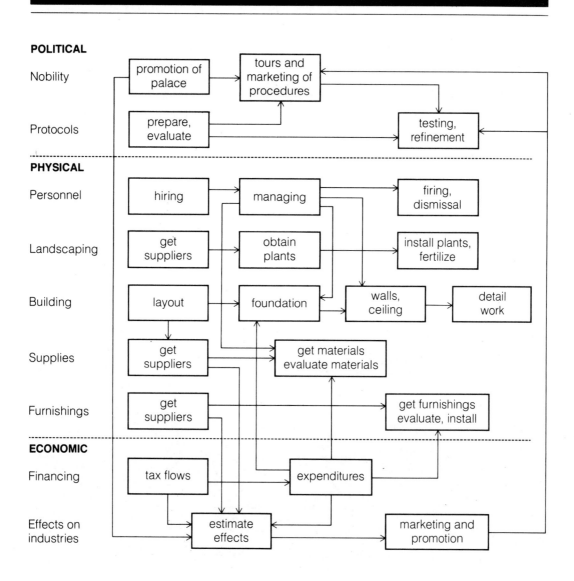

Building Stage Milestones

Once you've developed a plan and have completed its analysis, you must extrapolate a complete set of milestones. This is the first concrete step toward revising the project plan. Once you have a list of coordinated milestones, you must allocate resources (money, people, time, and equipment) in the way best suited to getting the job done.

These revised milestones (in detail) will represent your control document during building. With such a document, Colbert can stroll through the rising palace at Versailles and determine how segments of the building are progressing, which work crews will be needed next, and so on. He can also keep ahead of the budget requirements by finding out which segments are on schedule, behind schedule, ahead of schedule. The same goes for the less tangible—but more important—political and economic solutions, which are Louis's real goal.

Once you have completed the revision and expansion of the project plan, it's a good idea to schedule an approvals session with the user—especially on large projects. Explain to him, in his own language, what you expect to have completed, by when, and at what cost. Such an understanding between the

Analysis and development teems over the rising château. Louis sponsored in Versailles one of the greatest public works projects since the pyramids.

user and the system development team will prevent many misunderstandings during the course of construction.

Building Stage Activities

What goes on during the building stage? Obviously, there's construction. But there's more, part of it in aid of the construction, part of it looking backward, part looking forward to the next stage (installation). Generally speaking, we can say that the following activities must be completed during the building stage:

1. Construction: physical, paperwork, whatever.

2. Testing, which is continual: first of pieces, then of pieces fitted together, then of whole units, and so on.

3. Production coordination, which consists of quality control and administrative tasks such as budgetary revisions and scheduling.

4. Development of user training materials, so that the installation stage can be accomplished on schedule.

The aggrandizement of the Versailles gardens, with the château in the background. The building was criticized at the time because it "had no roof."

5. Documentation; the system development team is at all times answerable for its performance to the user.

In most cases, the building segment of those activities will be the most costly and the most consuming of personnel and time. For that reason, it usually commands the most attention. It's extremely important, however, that the other segments of activity be well attended to. Let's have a look for the moment at number four (development of user training materials).

User Training Materials

The dimensions of the task of developing user training materials vary considerably from project to project. In a simple system, the task may consist of simply noting for the user where the on/off switch is and what the term of a warranty is. In a complex system, it may include any or all of the following:

1. Operations manuals.

2. Repair and service manuals.

3. Procedures manuals.

4. Manuals to train the trainers.

5. Testing manuals.

6. Special materials such as circuit designs, floor plans, and special tools.

7. Anything else the user might need to learn to operate the system.

That can add up to quite a task, and it frequently does.

Unfortunately, preparing user training materials is a task that is often shoddily done—for no particular reason. Shoddiness can take various forms. A computer operations manual can be overwhelmingly (and mystifyingly) technical and undecodable. Procedures manuals can be too formal to be usable. A repair/service manual can be too disorganized to allow for troubleshooting by the user, and therefore unusable.

Too frequently, the task of developing training manuals for the user is either left until too late to be done right, or it's assigned to the least capable member of the system development team or—most horribly—to a member of the team who can't express himself in written language. What comes out of such assignments seldom makes sense.

There's no getting around the fact that training manuals are stepchildren. They are really of very little concern to the system development team (which is never going to use them), and the user doesn't know what he will need (and if he did, he probably couldn't develop it himself—he wouldn't be familiar with technical matters). So, the whole task gets a low priority and a low budget. But beware!

While the culprits shall remain nameless (and more common than you probably expect), there roam the world sophisticated system development teams such as the one whose whole thirty-million-dollar project was dumped because no user could figure out how to operate it. They stalk across oceans and across deserts seeking out users who are sucker enough to hire them. Consider the list below of systems that fail because of inadequate training manuals:

1. A child's toy that doesn't get put together on Christmas Eve because you simply can't figure out what fits into what.

2. An experimental word processing system, which failed because nobody could train secretaries to use it.

3. The 1976 Tax Reform Act, which few CPA's in the country could understand, and which was not explained either by Congress or by the Internal Revenue Service. (The same can be said of much legislation.)

4. James Joyce's novel *Finnegan's Wake,* for which no training material (key) has ever been found.

5. Medieval magic and witchcraft, because its training materials were kept secret, then lost.

6. Countless office filing systems, which are devised by individual secretaries and remain unintelligible to the rest of the world.

These examples should bring home some of the importance of clear, adequate, and accurate training manuals.

How Do You Know It's Going to Work?

There's an element of terror in building anything—an element of uncertainty. The kid who's putting together his first crystal radio set experiences it; the successful architect who's been asked to design a new cathedral feels it. Will what I'm building function the way it's supposed to?

Obviously, you wouldn't be building if you didn't have a reasonable expectation that the project would turn out well—at least one hopes you wouldn't. (With some systems it's hard to tell why someone didn't see the light sooner.) That's because during the design stage you (or someone else) worked out a design that would probably function correctly. The kid's crystal radio set was manufactured by a company that knows how to manufacture crystal radio sets, so the kid has faith that it will function if it's assembled correctly. The architect has included all the essential functions in the cathedral (an altar, pews, stained glass windows, a pulpit, easy access through doors and aisles, and high ceilings for intense quiet and good musical acoustics), so he has a reasonable expectation that things will work.

But is that where testing stops? At design? It better not! Whatever you have put together in the design stage is just a design. It may be based on tons of previous experience, on solid data from dependable sources, but it's just a design; it isn't a system that is ready to function. As any architect will tell you, there are always alterations during building to cope with unforeseen circumstances. How do you find those unforeseen circumstances? You look for them. How do you look for them? You perform tests on what you are doing.

The kid building the crystal radio tests each soldered wire to see that the soldering has really hooked the pieces together. The architect tests all kinds of things in the cathedral:

1. He tests the pews for comfort and beauty.

2. He tests the steps leading to the doors for ease of access.

3. He tests the pulpit for access, visibility from all points of the church, proper microphoning, light sources, etc.

4. He tests the altar against physical and ecclesiastical criteria.

5. He constantly inspects the whole building for ambience, harmony, quality, proportions, and so on.

In building a system, you test for two things:

1. How well the design is coping with the reality of building.

2. How well the design is being implemented by the builders.

Anytime there's a failure, one of the two areas above is malfunctioning. So your first question is obvious: Is the failure one of design, or one of building

Louis visits the Gobelins tapestry works, where Colbert is supervising a host of separate activities. Note the feeling of prosperity and affluence, compared to the economic sluggishness prevalent on Louis's accession to the throne. There is an apparent relationship in the composition to the contemporary Dutch Masters, but the style is unmistakably, self-indulgently, joyously French.

quality? If the answer to that question is that the builders are doing things wrong, the problem is relatively easy to solve (provided that they are doing something wrong on their own hook and not because you told them to do it that way). You either fire them and get someone who knows what he's doing, or you set your current builders on the straight and narrow again.

But what if the failure is really one of design? Suppose the architect begins to realize that for one reason or another a whole side of the cathedral will be insufficiently lighted? Suppose the computer programmer realizes that because something was not treated by the designers the security measures on the new system are easily broken? Suppose that a clinical psychologist realizes suddenly that the questions he's asking in his preliminary interviews are leading patients to the wrong answers?

Obviously, the answer is "back to the drawing board." And that's a costly answer, because we aren't in the design stage now; we're in the building stage. That means that either the workmen take a long break (a few months, in the case of some serious design problems), or you simply live with the error, or you try to design an enhancement that can be installed later. None of those alternatives is totally desirable.

The key to avoiding such problems is testing. Constant testing. Thorough testing. Varied testing. *You must find out what's going wrong (or right) as early as possible* to avoid excessive backtracking. The architect can't wait until the first service in the cathedral to find out that the people on the left side of the nave can't read their hymnals because of the darkness. Colbert can't wait until everyone is moved into Versailles to find out that the trip from the kitchen is so long that the food is cold by the time it is served. No indeed.

Types of Tests

There are basically two types of tests:

1. Tests of individual parts.

2. Tests of how individual parts fit together.

It would be simplistic to say that if all the parts of a crystal radio are not defective, you can simply assemble them and test the whole. You may very well go wrong that way, because when it doesn't work, you will have to take it all apart and start over.

You must test how each piece fits each other piece as you are putting it together. By so doing, you drastically reduce your chances for failure or delay because you isolate the problem as soon as there is a malfunction.

Figure 7.5 shows a progression of tests ranging from tests of individual parts to tests on the finished palace at Versailles. Note in figure 7.5 that we are testing in three arenas: physical, political, and economic. We have to test in all three, because everything we are doing has three sides:

1. Physical, because it is a physical palace we are building.

2. Political, because the group I and group II lists seek a political solution to Louis's problems.

3. Economic, because the group I and group II lists seek a solution that is economic as well.

The Queen's bedchamber. Is there enough gilt and crystal? (yes) Is it regal? (yes) Does the wallpaper harmonize with the firescreen? (mais, oui!)

Levels of Testing in the Palace of Versailles

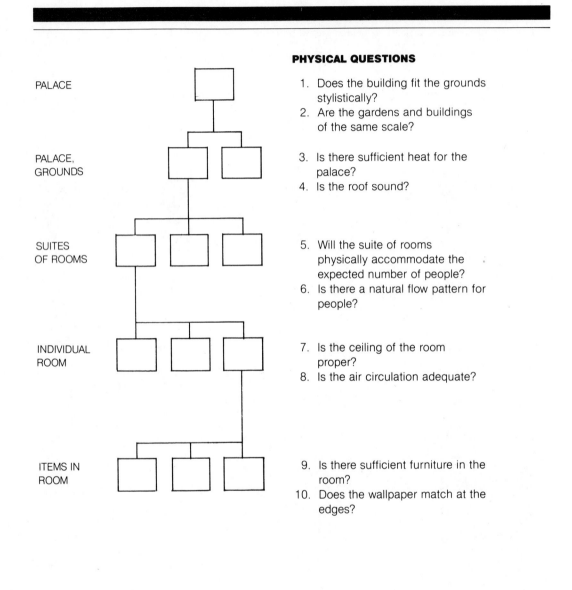

PHYSICAL QUESTIONS

PALACE

1. Does the building fit the grounds stylistically?
2. Are the gardens and buildings of the same scale?

PALACE,
GROUNDS

3. Is there sufficient heat for the palace?
4. Is the roof sound?

SUITES
OF ROOMS

5. Will the suite of rooms physically accommodate the expected number of people?
6. Is there a natural flow pattern for people?

INDIVIDUAL
ROOM

7. Is the ceiling of the room proper?
8. Is the air circulation adequate?

ITEMS IN
ROOM

9. Is there sufficient furniture in the room?
10. Does the wallpaper match at the edges?

POLITICAL QUESTIONS[1]

1. Does the nobility only have honorial prerogatives?
2. Is the center of government at Versailles?

3. Are the grounds and palace imposing to the nobility?
4. Do the suites for the higher nobles have better views than those for the lower nobles?

5. Are the rooms for the nobles separated sufficiently from the government offices?
6. Can each room accommodate the noble for semipermanent residence?

7. Is the bedroom for the higher noble bigger than that for the lesser noble?

8. Are the furnishings sufficiently personalized to make the residents feel at home?

ECONOMIC QUESTIONS[1]

1. Is the tax base firm and growing?
2. Is French industry well established?

3. Are the new French industries receiving other business from nobles and others?
4. Have French personnel and goods been used throughout the palace and grounds?

5. Has the money spent been recycled through the tax base?
6. Was the money spent on each suite proportional to the status of the users of the suites?

7. Does the overall appearance say "French" loudly enough for visiting dignitaries to want to buy French products for their own palaces in other countries?
8. Are the furnishings sufficiently stylish to ensure continued business to industry with changes in style and taste?

9. Were enough chairs ordered from different manufacturers to give prestige to the manufacturers?

1. These questions refer back to the Colbert group II list, Figure 4.8, page 87.

Likewise, you must test any system both in terms of physical functioning and in terms of the group I and group II lists. Those tests might be simplified to:

1. Are things functioning, singly and together?

2. Are they functioning in a manner that will transform the problematic environment into a sustaining environment?

Forms Tests May Take

Let's talk first about tests that can be performed on individual parts, because those tests are critical to the entire building stage. If the individual parts fail, the whole project fails. If the light switches don't work in the cathedral, then services at night are not possible. If the wire coil on the crystal radio set doesn't work, then there's no point trying to fit the other pieces together.

Individual parts tests are of several types, including:

1. *Logical Tests.* In a logical test, you anticipate probable use and test for it. For instance, if the queen's bedchamber must accommodate forty people comfortably for morning gossip fests, you examine the room to see if there's space for forty people. If an electric circuit will probably have to conduct enough power to run a

Holy Innocents Fountain, built by Louis XIV in Paris to ease a chronic shortage of water. Tests would have to be performed in two areas: is it beautiful? does it deliver sufficient water?

refrigerator, a washer, and a dryer, you check to see if that amount of power will actually flow through the lines.

2. *Limits Tests.* In a limits test, you anticipate a use considerably heavier than the probable use and test for it. You see if, for example, one hundred people can fit into the Queen's chambers. You see if a color television, a refrigerator, a freezer, a washer and dryer, an electric typewriter, and a photo darkroom can be adequately supplied through the existing circuitry. If they can, you note that test for the user's benefit.

3. *Failure Tests.* In a failure test, you continue to load use on a part until it fails. You run voltage through a piece of wire until it melts; you slam a car door until it falls off. Obviously not a test you use on parts of limited availability, failure testing is most valuable on parts which will have to be installed in various places: light switches, on/off buttons, and assembly line parts in factories, for example.

Measuring Test Results

There's no point to testing if the results are not tabulated in some fashion and used. If you test a wire coil on a crystal radio, find it to be defective, and proceed anyway, the time spent on the test was wasted.

As a larger example, if you're the cathedral architect we spoke of earlier, and you find the acoustics of the choir to be defective, you can't use that information until it's formalized in some fashion. You can't begin altering plans for the choir loft until you know what's wrong and how wrong it is. Once you find that the bass reverberations shake the stained glass windows (threatening to crack them), you can deal with the problem created by the defect: strengthen the windows or decrease the bass power of the pipe organ.

In order to make test results useful, you must adopt a measuring system to give accurate and usable data. The measurements you need will fall into two categories: *quantitative* and *qualitative*. The scales and units used in measurement will vary from test to test:

1. A temperature test will be measured in degrees.

2. A sound test will be measured in decibels.

3. A lighting test will be measured in candles.

4. An automotive power test will be measured in horsepower.

The order in which tests are made and measured is always the same:

1. Individual parts are tested individually (chairs, nails, water pipes, flooring oak, ovens, mirrors, and so on).

2. Tested individual parts are integrated into small units and tested for how well they work together.

3. Tested small units are integrated into larger units and tested for how well the larger units work together.

4. This process is repeated through the final series of tests, which measure how well the entire system works.

Figure 7.6 lists some of the types of questions that should be answered in testing individual parts of a system. Figure 7.7, in contrast, lists some of the questions that should be asked in testing integrated systems, whether they be small units or whole systems.

Obviously, we've made no attempt to be exhaustive in compiling these sample questions. The possible subjects of completed systems are almost

FIGURE 7.6

Measurements for Testing a Part of the System

QUANTITATIVE

1. How much does it cost to run?
2. What are the physical dimensions of the part?
3. What are testing boundaries of the part?
4. Under what quantitative conditions will the part fail (for example, temperature, occupancy, or volume)?
5. Is the part easily duplicated and replaced in terms of cost?
6. What is the capacity of the part (volume, quantity, and so on)?
7. Who can maintain the part if something goes wrong?

QUALITATIVE

1. Is the part beautiful?
2. Is the part easy to use?
3. Is the part easy to understand and maintain?
4. Is it likely that the part can be abused?
5. Does the part fit in with its anticipated surroundings?
6. Does the part have enough spare room or capacity to support other functions?
7. Does the part fit in with the user organization and tastes?

infinite; therefore, the range of tests is also nearly infinite. Nevertheless, the questions in figures 7.6 and 7.7 should point the way toward the types of tests which should be made. (It may be useful at this point to review figure 7.5 for application to the Versailles case study.)

Managing the Building Stage

The management role of the system development function lives or dies on how well it handles the activities of the building stage. Until now,

FIGURE 7.7

Measurements for Testing an Integrated System

QUANTITATIVE

1. How much does it cost to run?
2. Does the system satisfy the requirements as noted in the system specifications?
3. Under what conditions will the system fail? How will failure affect the user organization?
4. What is the capacity of the system?
5. What excess capacity, room, and other measures exist?

QUALITATIVE

1. Is the system beautiful?
2. Is the system easy to use?
3. Can the system be understood by the user?
4. What are the chances of failure?
5. What is the likelihood that changes will be requested? What will be time and effort to implement changes?
6. Does the system appear to satisfy the benefits the user will expect?
7. Can new users understand and work with the system easily?

management has been able to coordinate fairly tight operations, always under the approval of the user. He has worked with a feasibility team (A&D), a group of project analysts, and a team of designers. He has appointed project leaders who are competent and with whom he works well. But until now he has not had to coordinate anything beyond data and designs, and he has always been able to fall back on user approval.

The administrative tasks that fall to management at this point are all-important:

1. Scheduling, budgeting, and general adherence to the project plan.

2. Coordination between various project leaders where needed.

3. Accounting and accountability.

4. Evaluating important test results and initiating remedial action.

5. Ensuring proper documentation and files organization.

But there's a more important task as well: the revision of user requirements during times of crisis. In most building stages, you'll find that some user requirements must be altered or done away with entirely and unexpectedly. The strategic decisions to be made in connection with these issues are the weightiest decisions that management will have to make. In some cases, the user may be contacted for approval; in others, decisions will have to be made on the spot. The two examples that follow illustrate the possibilities:

Example 1.

We know there is a user requirement that all components of the palace be of French manufacture. Suppose that we encounter an important component that for one reason or another isn't and can't be made in France. A good example of such a component would be marble for the facade; it must be quarried in Italy. Management's (Colbert's) dilemma is that no matter what decision he makes, he's going to be contradicting a direct order from Louis. If he opts for French granite, he won't achieve the look of exterior elegance that Louis feels is totally necessary to the success of the project. If he decides to go to Italy for marble, he violates the rule about using French materials. If he decides to halt construction until marble is found in France, he will delay scheduling and increase the allotted budget. What is he to do? Obviously, he'll confer with the user (Louis) for a decision.

Example 2.

Colbert finds that the fountains in the gardens may not get an adequate water supply from the existing wells; therefore water will have to be found elsewhere. Any search for a new water supply will be costly and time-consuming and will interrupt the smooth flow dictated by the approved project plan. Nevertheless, Louis is a busy man with little time for such decisions, so Colbert must decide on his own to dig new wells, spend unexpected money, and delay construction to whatever extent is necessary.

Strategic decisions during building are those based upon changes in the user requirements or in the approved project plan.

Managerial decisions are those that require minor alteration of the project plan or that are involved with administering the project plan: at what point during construction can the fountains near the palace be installed without danger of being trampled by workmen?

Tactical decisions are on-the-spot decisions required by emergency situations: Water begins to seep into the foundation trenches from a previously unknown artesian source. How can that seepage be stopped before the foundation trenches are destroyed?

The Importance of Communication

There's no way that management can adequately control the building stage without a good communications network. His project leaders must report all important details to him immediately and completely, because he is the custodian of the overall project plan. Only management can solve problems created by unexpected delays that involve more than one project leader. Only management can authorize new budget expenditures.

Unless management is in constant contact with those below, and only if the project leaders are constantly informed of work progress and problems, can a building stage be brought to fruition. This flow of information must be unobstructed. Refer to figure 6.8 (page 134) for a summary of the interfaces that must be considered in this communications network.

The success of the entire project may well be dependent upon timely and correct information. Enough said.

The move to the installation stage

At the end of the building stage, you must be totally ready to begin turning the system over to the user. That means that *everything* must be adequately tested; that the training materials must be in order; and that the system paperwork is complete for a check for accountability.

Once you move to installation, there's no turning back. Any defects that must be corrected after installation are more expensive than ever (as indicated in figure 7.1).

When you are totally satisfied that the system is ready for use, you are ready to get approval from the user to proceed to the next stage. Ready? Set? Go!

Aerial view of a portion of the château.

Exercises

1. Define the following terms:
   a. System testing.
   b. Types of testing.
   c. User procedures.
   d. Operations procedures.
   e. System integration.
   f. Training materials.

In each of the systems in problems 2 through 6, describe what could go wrong in the building stage that *(a)* is due to faulty logical design, *(b)* is due to faulty physical design, *(c)* is due to problems with the system specifications, *(d)* is due to the construction itself.

2. The building of a house.
3. The construction of a telephone book.
4. The development of the electric light bulb.
5. The administration of a questionnaire on gambling patterns in Las Vegas.
6. The integration of a school district by cross-town busing.

In each of problems 7 through 10 discuss how you would conduct system testing. Where would you collect data for the test? How would you assess the results of the test?

7. A new method for eliminating mosquitoes.
8. An automobile that runs on a mixture of gas and water.
9. A company telephone directory with emergency directions in case of fire or other disaster.
10. A traffic light at a corner that previously had stop signs.

In problems 11 through 15, construct an outline for a manual to be used in working with the system.

11. A record player.
12. An air-conditioning system.
13. A payroll system.
14. A report on school busing.
15. An approach to reduce alcoholism.

System failure can occur simply because of a lack of communication between workers and managers who are trying to build a system. Discuss possible sources of problems in cases 16 through 20.

16. The building of a house.
17. The construction of a tunnel under a river.
18. The preparation of a survey on drivers' attitudes toward motorcycles.
19. The filing system for periodicals in a library.
20. The layout of a loading dock at the rear of a plant.

21. You work in a department that makes the fenders for two-door sedans. You are assigned to construct a system to ensure that the fenders are properly made. You have been given a week to build and install the system. If you don't have the time to develop a detailed design, what do you do? What are the advantages to not planning or developing a design?

22. In the late 1960s and early 1970s, the State of California built a new governor's mansion. The old one was not acceptable. After the new one was built, the new governor refused to live in it. He claimed that to use it would be more expensive than letting it lie idle. Examine the causes of failure even though the mansion is thought by some to be an architectural and construction success.

In each of the cases in problems 23 through 26, discuss possible reasons for a system to be over budget and behind schedule at the end of the construction activity.

23. The building of a house.
24. The construction of an advertising campaign on a new deodorant.
25. The conduct of a survey to determine the relation of income level and type of automobile driven.
26. The construction of a payroll system.

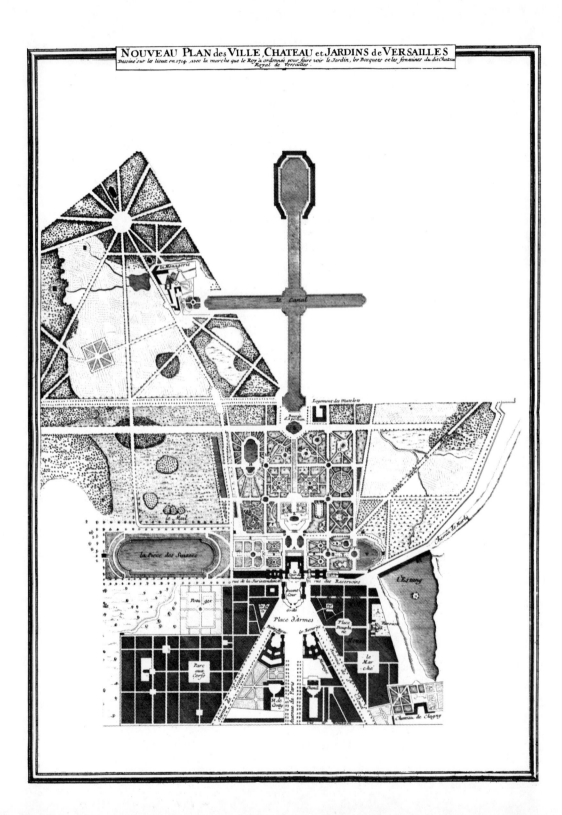

NOUVEAU PLAN des VILLE, CHATEAU et JARDINS de VERSAILLES

Dessiné sur les lieux en 1714, avec la marche que le Roy a ordonnée pour faire voir le Jardin, les Bosquets et les fontaines du dit Chateau Royal de Versailles.

8

The Installation Stage

━━━━━━━━━

Before the royal inspection party arrives, you stand in the enormous courtyard in front of the finished château and admire your work. From the distance of three centuries, we know it to be one of the monumental artistic productions of Western civilization. You feel the grandeur, power, and pinnacles of taste and luxury just from staring at it. You say to yourself, "Well, it's finished."

Heh, heh, heh. It might be nice to say to yourself then, but it isn't the case. As you may observe from the fact that there are four more chapters in this book, few systems are finished when construction has stopped.

What Happens in Installation?

For the purpose of simplification, we can divide this stage into two main components:

1. User acceptance testing (and consequent modification).

2. User installation.

169

A common temptation in system development groups is to skim over the installation stage after a pass at user training and call the whole thing quits much too early. That courts failure, as you will see later on in this chapter. A user "trial by fire" is a yes or no situation—all or nothing. If you've designed and built a payroll system, and your instruction to the user is, "start living with it," you risk having the whole project thrown back in your face at the first misfire.

User Acceptance Testing

You hear the clatter of hoofs on the cobbled streets of the village at Versailles. It's the royal party. The king arrives to inspect Colbert's project. Here's what

An early view of the château with a good vista of the gardens. The building eventually grew to match the splendor of the landscaping, not vice versa.

Colbert may expect:

1. An entirely qualitative initial reaction. ("How pretty is it? I thought it would look different.")

2. A difficult birth process, with a protracted labor on your part. ("Thank you for what you have done; you may leave now; wait a minute, I don't like the tapestries, furniture, fountains, flowers, carpets. . . .")

3. Ignorance (total or partial) on the user's part as to how the system operates. ("Where is the front door?")

A view of the California Governor's Mansion constructed under the Reagan administration. The first governor who could have lived in it, Jerry Brown, rejected it outright because it was too large and expensive to maintain. Different user requirements strike again.

4. Surprise from the user wherever you were forced to alter his requirements in the building stage. ("I thought we were going to have a marble facade.")

5. Impatience. ("Bring me some food. What do you mean there isn't any food here yet?")

And all of that is justifiable on the user's part. Remember that he wasn't much involved in the building stage; the last he saw was a set of blueprints that he didn't really understand anyway. Now he sees a finished palace. It's like reading a novel and then seeing the movie: the actors never look the image of the character you imagined while you were reading. You've got to give the user a chance to adjust before you try to do anything else with him.

In a modern example, that period of adjustment may include any or all of the following:

1. Learning to live with the final cost.

2. Assessing the impact of installation on current operations.

3. Reconciling immediately discernible differences with what he expected.

4. Conditioning his organization to be ready to change over to the new system.

5. Learning to live with the fact that operation of the system isn't yet a fact.

One of the system development team's primary responsibilities at the beginning of the installation stage is to keep the user from jumping the gun. Almost every user will want to wade right out into the water and start swimming in his or her new pool. It's your responsibility to inform him that the heater hasn't been turned on; that the place he's getting ready to walk into is the deep end; that there are only two sets of steps leading out of the pool and where they're located. Which brings us to the issue of user training, the first step in user acceptance testing.

User Training

Sat celeriter fieri quidquid fiat satis bene. Got that? It's an important concept, absolutely essential to properly accomplishing the work of the installation stage, as it is, of course, to the rest of the life cycle. It's—what do you mean, you don't understand it? It's perfectly clear Latin! Read it again; maybe it'll be clearer.

Enough is enough. The reason for pulling such a trick on you is that you need to know how the user will feel when you set the A&D team on him to explain the workings of the new system in their technical jargon. The time spent may be totally useless. Some people reading this text may be fluent Latinists, and will have no trouble with *sat celeriter fieri quidquid fiat satis bene.* Odds are, though, since Latin isn't part of the regular university curriculum, that the percentage of Latinists reading this text will be quite small.

The same goes for users. Odds are, since they needed to call in specialists in your field in the first place, that they don't speak (at least not fluently) the technical language of your profession. The point is, *you must speak to the user at all times in language that is clear to the user.*

In case you're interested, the Latin proverb quoted above is attributed to Caesar Augustus, and means (freely translated), "A project is finished quickly enough when it is finished well enough." Good advice to system development teams, don't you think?

Another thing that must be made absolutely clear before we get into training is the purpose of training: *The purpose of user training is to inform the user how to operate the system.* That purpose does not include justification of how the system was built nor self-aggrandizement of the system development team. In other words, don't use the user training phase to make the system development team seem more important than it is. Don't try to impress anyone; the user will be sufficiently impressed if you have done your work well. After all, that's what he hired you for. The only thrust of user training is getting information to the user on the operation of the system.

Refer to page 152, chapter 7, for a list of the types of user training materials that may have been constructed during the building stage. Those materials will now begin to function as both training and reference tools for the user. In that sense, they're the first parts of the system to become operational. The importance of that realization is enormous.

The first real indicator that the user will have of your total competence is the quality of the training materials you have put together. If those training materials are of poor quality, disorganized, incomplete, or unintelligible to him, that will color the user's perception of the entire system. In other words, if you can't tell Louis in words he can understand where to find the nearest chamber pot, he may well wonder whether you know what's going on in the system you built.

Take care with your training materials, and take care with user training. It's a valuable time for both user and system development team. It's the first indication for you of whether the system will need modification. Recall the systems we mentioned earlier (page 153) that failed because training was not carried out thoroughly or intelligibly.

The point you are trying to get to with user training is simply this: you must inform the user about the operation of the system well enough for him to test the system prior to acceptance. Louis can't test the chamber pot until he finds the chamber pot and learns where the secret latch that holds the lid down is. Then, if it leaks, well, you can fix it.

Top Down Versus Bottom Up Training

As with design, there are two basic approaches to user training: top down and bottom up. Each has its virtues and its disadvantages. Top down training begins with and continues to concentrate on the larger picture, the overview. It attempts to coordinate all the small pieces into a total system, with the total system being the primary focus.

The advantages of top down training should be immediately obvious. When successful, it provides two essential things:

1. A framework for specific training on subsystems.

2. A functional overview of the entire system from an administrative, financial, or production viewpoint.

Successful top down training integrates the system from a deductive position. The user learns the overall principles, then deduces the particulars from the general.

The disadvantages of the top down training approach aren't as immediately discernible. Largely, they are what might be termed "human failures" and are based on the simple fact that the user may not be able to comprehend the overview picture as quickly or as thoroughly as you might hope.

The method used in this textbook is top down. Recall that we discussed the shape of the life cycle in chapter 1. Since then, we've dissected the individual stages one by one, in the hope that the organizing discussion in chapter 1 would provide an adequate frame of reference for the student. If we're successful, the student will finish the course with an excellent overview of the systems approach and a good enough understanding of the particulars to know, at a minimum, where to look to find the details that don't spring immediately to mind.

Failures in top down training occur for one or both of two reasons:

1. You haven't adequately explained the overview, and none of the specific information "fits together." In the case of this text, if you didn't understand the discussion of life cycles in chapter 1, you won't be able to understand the importance of the staged approach, which is the organizational basis of the rest of the book.

2. You haven't adequately simplified the overview for your initial explanation. It requires too much technical detail for the user to understand it. He gives up, or pretends to understand concepts he cannot comprehend, or thinks he understands because you haven't been sufficiently clear.

These are insidious failures because they don't become immediately apparent. They do, however, become apparent when the user tries to implement his training. When he knows how to operate each subsystem, but doesn't know which comes first, second, and third, you may be sure that he doesn't understand the overview.

Bottom up training approaches the system from exactly the opposite viewpoint. The basic principle of bottom up training is that hands-on training orients better than theoretical training. In the Versailles case study, bottom up training is used to show Louis how his personal apartments are organized. Instead of telling him the philosophy behind the organization and the fact that the rooms are based on the shape of a rosette, Colbert shows him immediately where the bed is, where the chamber pot is, and where the secret staircase is. Then, once he knows the particulars, Colbert points out the larger design.

Successful user training usually incorporates some of each of these techniques. The mix is best determined by the complexity and the technical requirements of the system, and by the technical competence of the user and his staff.

Classroom Techniques in User Training

There are four basic tools that the system development team may use in user training:

1. Formal lectures, indispensable in top down training, provide administrative overview, philosophical overview, and so on.

2. Informal discussion.

3. Case studies handled in a walk-through fashion.

4. Hands-on training, which is the starting place for bottom up training.

Keep in mind that many systems will require a period of *operations training* as well. In installing a word processing system, for example, you must train the typists to operate the equipment in addition to training management to administer the system. Operations training teaches the Versailles gardeners how to care for the tropical plants in the Orangerie; it shows the servants how to find the back staircases; and it shows the cooks how to stoke the ovens in the kitchen. Operations training must usually be accomplished in a bottom up mode, since operations personnel are usually not involved in the whole system but remain confined to operating a single subsystem.

The User Tests for Acceptance

Once the user is fully trained in the operation of the system, he must carry out his own acceptance testing. These tests aren't the same as the ones you carried out at the end of the building stage because the user is now testing from his own proprietary viewpoint.

Familiar examples of user acceptance testing are myriad; each one of us has performed such tests:

1. New automobile purchase. The prospective buyer test-drives the automobile. He checks to see that his luggage will fit into the trunk, that the backseat has enough legroom for his Uncle Bill, that the sun visor is adequate for his five-foot wife, that the FM radio has stereophonic sound, that the car fits into his garage. These tests aren't similar to the tests that the manufacturer performed—how could they be? The manufacturer doesn't know what size the garage is, nor what size Uncle Bill is.

2. An automated payroll system. A system development group has developed an automated payroll system to replace a manual system for company X. It is fully tested to the most recent user requirements; failure tests have been performed to tell the user how much data can be handled, at what speed, and with what accuracy. User acceptance testing in this case, however, revolves around completely separate issues: Can the user's current staff learn to operate it? Can the existing telephone lines support the number of data sets required? Will the building have to be rewired to provide dedicated electrical currents? Are the ink-spitting line printers legible enough to be read by elderly retirees?

The Orangerie at Versailles. Louis would check it for beauty, taste, and accessibility. He would not check it for practicality: Louis wanted, and got, fresh oranges year round. It was of little concern to him that all 500 trees would have to be protected from snow all winter.

If you are a member of a system development team, your responsibility during user acceptance testing is twofold:

1. Assist the user wherever necessary in performing the tests.

2. Don't assist the user too much, or the tests may be invalid.

These two responsibilities may seem to be contradictory, but they aren't. A fine line separates the two. If we, as system developers, sit down with the user and instruct him carefully in the operation of a new computer terminal, then assist him in his first run, we are probably within the boundaries of proper assistance. If, however, we find that the user cannot operate the terminal, and we sit down and do it for him, the test is invalid. We won't be there as members of the user's staff, so such testing by the user depends on a false element: us.

Modification and Alteration

Very few systems escape some modification during user acceptance testing. Not many people buy a showroom model of a car without asking for some additional equipment (different color, different radio, radial tires, wheel covers, different upholstery, or different transmission). The business world doesn't contain a payroll system that was installed without incident and modification. That just isn't the nature of things.

The user will discover things about the system that you either never dreamed of or never considered. You may never have known that retirees would be getting their benefits statements from your payroll system—and that your ink-spitting line printer might cause legibility problems. But the user (who has to answer to the government about his benefits administration) will be very aware of that problem. The probable solution is a different model line printer.

Whatever the alterations are, reconcile yourself to the fact that there will be some. Figure 8.1 demonstrates the relationship between the group I list and user requirements and the extent of modifications at the point of installation. Notice that the more complete and accurate the group I list, the fewer the alterations during installation.

FIGURE 8.1

Effect of Modifications During Installation of a House

GROUP I LIST **USER REQUIREMENTS**

C: Four people in family Three-bedroom home

F: Three bedrooms minimum

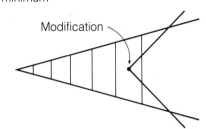

Total development cost = development costs through installation + cost of redesign + cost of rebuild + cost of reinstallation

GROUP I LIST **USER REQUIREMENTS**

C: Four people in family; in two or three years, add one more Four-bedroom home

F: Four-bedroom home

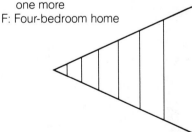

Total development cost = sum of stage costs through installation

The effects of not anticipating the addition of a fifth member to the family.

The Cost of Modification

Back to figure 7.1 (page 144). Remember when we said that cumulative costs escalate without pause throughout the system life cycle? It's true. If we have to delay installation now until we can select and install a new line printer for the payroll system, we will spend much more time and money on that line printer than we would have if we had foreseen the problem. We may have a whole system waiting for one component.

Overall costs of modification probably won't be larger than overall construction costs (at least we hope that's the case). But we still have some way to go during this stage. After all, we called this stage installation, and we haven't begun to install the system.

Acceptance Testing at Versailles

We have discussed previously the three different realms in which the Versailles system is expected to operate:

1. Physically. The château is expected to provide an elegant, highly visible, luxurious home and life-style for the most powerful monarch in the world.

This enormous bas-relief medallion of Louis XIV in the style of Mars dominates Versailles' Salon de la Guerre. It is typical of the personal identification Louis insisted on throughout the château.

2. **Politically.** Versailles is expected to impoverish the nobility, make it dependent upon the crown, make the importance of the nobility entirely honorific.

3. **Economically.** The château is expected to revive French industry, channel it, enlarge Louis's tax base, and establish France as a world leader in fine manufacturing.

Louis's acceptance testing has to operate in all three realms. Recall that figure 7.5 (page 158) detailed some of the tests that the system development team, under Colbert, performed prior to the end of the building stage. Those tests were as complete and exhaustive as Colbert could devise. But Louis's tests are another thing entirely.

Louis will test for *actual results* rather than probable results. Figure 8.2 details some of Louis's probable tests during his acceptance testing period. Notice that, looked at together, the thrust of Louis's testing is different from the thrust of Colbert's testing during building. Louis's testing is of a living piece of machinery—of a system expected to be fully operational. Colbert's building tests were simply expanded versions of the logical, limits, and failure tests we discussed in chapter 7.

There are parallels to Louis's testing in modern business and industry, and the key to the difference lies in the fact that these are all tests of actual results as opposed to simple capabilities. Figure 8.3 demonstrates the differences in building testing and user acceptance testing for three modern systems.

Chairman Mao Tse-tung made widespread use of a personal identification program in the People's Republic of China. This poster shows the god-like presence of the Chairman directing the lives of youthful revolutionaries. The poster reads, "Follow the party line in the thought of Chairman Mao": China's counterpart to Louis XIV's personality cult in 17th-century France.

FIGURE 8.2

Louis XIV's User Acceptance Testing at Versailles

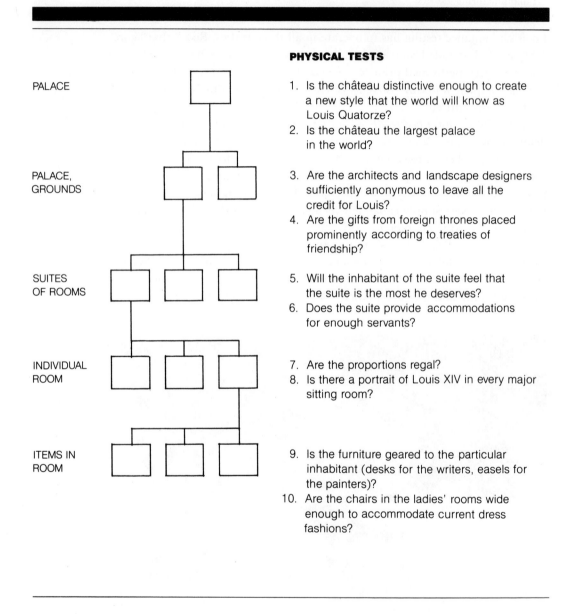

PHYSICAL TESTS

1. Is the château distinctive enough to create a new style that the world will know as Louis Quatorze?
2. Is the château the largest palace in the world?
3. Are the architects and landscape designers sufficiently anonymous to leave all the credit for Louis?
4. Are the gifts from foreign thrones placed prominently according to treaties of friendship?
5. Will the inhabitant of the suite feel that the suite is the most he deserves?
6. Does the suite provide accommodations for enough servants?
7. Are the proportions regal?
8. Is there a portrait of Louis XIV in every major sitting room?
9. Is the furniture geared to the particular inhabitant (desks for the writers, easels for the painters)?
10. Are the chairs in the ladies' rooms wide enough to accommodate current dress fashions?

Left labels: PALACE; PALACE, GROUNDS; SUITES OF ROOMS; INDIVIDUAL ROOM; ITEMS IN ROOM

POLITICAL TESTS

1. Is the château grander than the palace of any noble?
2. Do the highest-ranking nobles have the finest quarters?

3. Does the arrangement of rooms accord with the latest protocol?
4. Is the hunting preserve stocked with sufficiently varied game?

5. Does the style of the suite dictate a sufficiently large budget for the inhabitant (to put him close to bankruptcy)?

6. Will the rooms of wealthy nobles accommodate large gambling parties?

7. Does the furniture fit with the rank?
8. Are the inhabitant's own possessions sufficiently subordinated to Louis's gifts?

ECONOMIC TESTS

1. Is the tax base large enough and sound enough to support Louis's budget?
2. Will the maintenance stage employ enough workmen to ease the massive unemployment in the state?

3. Will foreign dignitaries want to emulate the style in their homes?
4. Will individual industries have to gear down now that the château is complete?

5. Are all the styles distinctively French to increase trade prestige?
6. Can more works of French art and industry be fitted in?

7. Are the artists and suppliers capable of maintaining the suite?
8. Will the style of the room encourage the inhabitant to further patronize the artists and suppliers?

9. Can maintenance costs be increased at the expense of the inhabitants to further support the tax base?
10. Does each piece of furniture make a statement about the quality of French inventiveness and style?

FIGURE 8.3

Different Tests During Building and Installation

EXAMPLE 1. **A HOUSE.**

A family looks around at various housing developments. The builder in each case has conducted some tests during the actual construction of the house. The family decides on a particular design, and after construction they arrive and test out the house themselves before signing final papers.

BUILDING TESTS:
1. Heating.
2. Functioning of the kitchen appliances.
3. Bedroom lights.

ACCEPTANCE TESTS:
1. Is the room warm and friendly?
2. Is the kitchen easy to work in?
3. Is there enough light to read with? Does the light disturb someone trying to sleep?

EXAMPLE 2. **TAX LEGISLATION.**

A legislative body has enacted tax legislation for implementation and use by the executive branch of the government. The executive branch asks several questions constituting the acceptance testing.

BUILDING TESTS:
1. Is the legislation internally consistent?
2. What will be the costs of collecting the new taxes?

ACCEPTANCE TESTS:
1. How many new loopholes are created by shortage of funds for enforcement?
2. Who is sufficiently competent to head up the tax collection?

EXAMPLE 3. **URBAN RENEWAL.**

The user here is the body of local government officials who must implement the decisions and plans of other groups.

BUILDING TESTS:
1. Estimated number of jobs created.
2. Anticipated traffic patterns.

ACCEPTANCE TESTS:
1. What is the estimated number of jobs displaced due to construction?
2. How many more traffic patrol people will be needed during operation?

Tentative Acceptance: The Beginning of Installation

Once user acceptance testing is finished and all modifications have been made to the user's specifications, the user will give a tentative go-ahead for installation. It's tentative because the user still hasn't seen the system *in operation*, however extensive the testing period was. He still has the option of chucking the system and starting over. In any case he will have to bear the brunt of costs he has authorized to this point, but you are by no means ensured yet that he has accepted the system as his own.

Nevertheless, the tentative acceptance is a good sign in a well-administered system development life cycle that things are finally going to come to fruition. After all, if Louis is finally going to take the big step—move into Versailles—you have a good idea that he won't order it dismantled tomorrow. With a man like Louis, however, nothing is out of the question. Colbert must have held his breath on the day that the regal procession set out from the Louvre for the new abode.

Types of Installation

Well, there you sit. The user has tentatively accepted your completed system; you've felt secure enough to turn over operation of the system to the user. What next? Does the user just move in and take over? Maybe. And maybe not. Consider what might happen in the following cases:

1. An automated payroll system. You've now solved all the problems mentioned earlier and have come to a point in system development where you feel the only possible next step is user installation. He has 1,500 employees who will be depending on your system for paychecks, and you still have not seen the system in actual operation. Do you trust in powers above and just recommend that the user scrap the old manual system and start using the new one? What happens if there is a snag, and the old system is no longer available as a backup?

2. Louis XIV has approved your new plan for reorganizing the Parisian court system. You haven't tried any cases in it, nor have you tested the appeals procedure (because you haven't tried any cases, obviously). But you can't open it provisionally, because that would leave Paris with two competing court systems (the

new one, and the old one operating as a backup). Who would have the last word on appeals? Who would decide on the jurisdictions?

3. You've been transferred with your company to a new location. You've been trying to sell your old house, but no buyers have yet materialized. However, you've found a new place you'd like to buy in the new city. What do you do? Pass up the new place until you sell the old—and risk the new being sold to someone else? Buy the new place and pay on *two* mortgages at the same time? Lower the asking price of the old house and cross your fingers? Rent the old house out, and face the problems of being an absentee landlord?

4. It's zero degrees Fahrenheit outside, and today is the day the workmen have finished installing your new solar heating unit. You have a four-month-old baby and a houseful of tropical plants that cost you a bundle. Do you turn off the gas furnace in the basement and trust that the solar units will be able to keep you warm? Do you operate both at the same time? Do you wait until spring?

As you are probably beginning to gather, installation isn't always a simple matter of getting in and starting to drive—not from your standpoint as system developer nor from the user's standpoint as possible sufferer in case of nonfunctioning systems.

There has to be more than one type of installation to cover the situations mentioned above. Relax. There are three basic modes of handling installation—and an almost infinite variety of combinations of those three.

This house in Davis, California, is being converted to solar energy. Since it is already inhabited during the conversion, installation will probably be on a parallel basis—electricity and natural gas service will be continued at least until the solar generator proves itself.

Parallel Installation

Parallel installation is the case where you operate both systems (the old and the new) at the same time until the new system is sufficiently operable and proven to stand on its own. There are two things that can be said immediately about this mode:

1. It's the safest mode. If anything goes wrong, you've got the old system standing there as a backup.

2. It's the most expensive mode. You're paying for two whole systems. In the case of a system involving personnel, for example, you're paying for two full staffs. Whew!

Despite its expense, parallel installation may well be the only real choice in some cases. Case four—the solar heating unit—seems to be a prime choice for parallel installation. You might turn on the solar heat in some rooms, the furnace heat in others (the more critical areas). Or you might turn them both on at half-capacity and gradually decrease the furnace heat as the solar heat proved itself adequate. You're going to be paying for both systems, but the consequences to the baby and the plants of possible failure of the new system warrant the extra expense.

There are distinct disadvantages to parallel installation in the area of personnel. If you decide, for example, to operate a new manufacturing facility in tandem with an old one producing the same product, you have a real training problem:

1. Do you move the old staff to the new facility, train them to operate the new equipment, and replace them with temporary workers at the old facility (who would also have to be trained)? Wouldn't the newness of the working staff at the old facility decrease the dependability of the old facility as a backup?

2. Do you hire temporary workers to operate the new facility and thus face the training period (and its associated expense) twice?

3. Do you hire new workers for the new facility and face the fact that the old workers will have to be laid off when the old plant shuts down?

4. Do you try a combination of the above alternatives?

These questions cannot be answered in generalities; they can only be answered after a careful analysis of the situation you encounter when your system is ready for installation. You might refer back to the discussion of cost/benefit analyses in figures 4.4 and 4.5 (pages 74 and 78) when making such a decision.

Pilot Installation

In a pilot installation, you operate the new system in a limited capacity, where you can remedy any failure by a quick withdrawal or a quick cover-up. The pilot installation mode is really a subdivision of the parallel installation mode, but is not quite as drastic. You don't have to operate two whole systems in a pilot; you have to operate the old system in a slightly diminished capacity and the new one in a strictly limited capacity. The pilot installation mode offers several advantages:

1. You can train old personnel to operate the pilot without having to maintain two staffs.

2. You can reconvert to the old system at the first serious failure in the new system.

3. You can work out problems in the new system on the job at minimum cost.

It has corresponding disadvantages:

1. You aren't really testing the new system as a whole. What works in the pilot is not guaranteed to work in the whole system. In the solar heating example, you might pilot-heat the bathrooms and then find that the conduit work to the baby's room is leaking.

2. You're still (let's face it) operating two systems, with no real commitment to either one. The staff at the new system isn't assured that they will eventually be working there permanently, and they may not try as hard as they should to make things work.

3. You're still incurring the costs of two systems, although not as heavily as in the parallel mode. You must maintain two administrative staffs (twice the red tape), separate budgets, an ongoing training program as you slowly convert to the new system, and so on.

The automated payroll system may be an ideal candidate for pilot installation. You could decide to handle the payroll and benefits programs of only one department on the new system. Then, if there's a failure, only one department is affected, and the failure can probably be covered by the old system in a pinch.

Sooner or later, though, you're going to have to get off the fence with a pilot installation. And when you do, you're going to be taking some risks. In the payroll system example, you're not going to find out if the system will overload too quickly until you approach its functional limits (which you're not going to do with a pilot). Still, you may be better off with pilot installation in many cases where functioning of the new system is just too important to risk a whole hog approach. However, be aware of its limitations, and be on guard for new developments even after a successful pilot installation has been completed and approved.

Immediate Installation

Last and most common is immediate installation. You buy a car, you get in it, and you drive it home. You install a new freezer in the kitchen and put a lot of meat and chicken into it right away. You elect a new government, and it goes to Washington and takes over.

The advantages of immediate installation are obvious:

1. It's fast.

2. It's the least costly if no failure occurs.

The disadvantages are equally obvious:

1. If a failure occurs, you're up the creek without a paddle.

2. You've got to train everybody involved overnight, and they take over with no experience as a team working on the new system.

With these things in mind, it seems to be the only choice in some situations. Usually these situations are those where the old system is so valueless that even failure is better. If your old refrigerator dies a rickety death, the new refrigerator will be an improvement, even if it overloads the fuses and you have to bring in a new power line as a result (after all, that's only a temporary failure; the old refrigerator was a total loss).

Failure

Getting used to the sound of that word? It still comes as a shock, doesn't it? Who would think that failure was even possible at this late stage in the game? If you think failure is unlikely or impossible now on a really grand scale, look into the BART transportation system in San Francisco or the Edsel automobile. Then you might reconsider.

Failure is always possible if you haven't done your homework. BART has never worked out its technical problems; Ford was entirely mistaken in its market analysis of the probable reception of the Edsel.

But there's a more insidious approach to failure in installation. It's what we might call *midinstallation paralysis.*

Midinstallation Paralysis

Say you're the commissioner of airports at a large American city. You've worked like the dickens over the past twenty years on getting approvals to build a new airport to serve the area, and it's a vast improvement over the outdated old metropolitan international airport. It's operational now, and you—wise planner that you are—decide on a parallel installation. After all, you can't risk the economic disaster that immediate installation might imply if the new airport were found to be defective in any of thousands of ways.

You've been working both airports in a parallel mode now for a year, and you're satisfied that the old one is ready to be shut down. The new one can handle all the traffic now.

But when push comes to shove, you find that the hotel conglomerate that has a corner on the accommodations market at the old airport is just not quite ready to relinquish its elegant (and profitable) facilities at the old airport. You point out the logic of the new airport to them; they point out their potential dollar loss to you—and to the city council, which depends on their tax dollars for school support and so on.

The city council can't make up its mind, and you end up having to operate two airports indefinitely. Therefore, airport district revenues will have to be doubled in the near future, and the city council will have to raise the sales tax by one cent. They blame you for poor administration when they run for office. You find yourself in a press conference trying to explain why you built this new airport when the city couldn't afford to operate it. You're forced to resign, and the city is left with two airports—neither operating at capacity. They're also left with two freeway systems leading to the airports. The hotel conglomerate finds itself forced to operate its hotels on half the previous number of room rentals; they lose money, and the tax base decreases. What can happen? The obvious. The city can close the new airport.

Midinstallation paralysis is common—much more common than you might expect. And it isn't common only in government. Beware!

Turning It Over to Maintenance

Somehow you manage to get your system installed (you clever devil!), and you're ready to pack up your bags and beat it. The user is in full command of the system now, and you're beginning to feel unemployed. It's all over, right? Wrong. As you'll see in the next two chapters, there's one more stage that the system will pass through: operation. And, although most traditional texts on system development would have you quit now, you must resist that temptation.

Try on this statistic for sheer stun value: for some systems, the operation stage accounts for as much as 70 percent of the life cycle costs. Sufficiently motivated to read further? You should be. But we have one more task to perform in this stage: a project summary.

Project summary

You may have noticed by now that we haven't been very demanding in this text about formal documentation. We haven't specified the numbers of formal reports usually required by other texts. Consider that fact when we now tell you that a project summary is an essential component of the life cycle. A project summary should enumerate everything that has happened during the development of the system solution. It might take as a jumping-off place the project plan—how well it worked, where the slipups were.

However you choose to organize it, the project summary must be adequate to serve two very important purposes:

1. It should be a learning document for your system development team, helping you avoid making the same mistakes all over again on the next project you tackle.

2. It should be a history of the project which will provide accurate and adequate information on whatever subject might be needed during the maintenance and enhancement stages.

We won't dwell on the project summary or its format too long (as with other documentation, the format and the extent of the summary will vary from industry to industry, with local conventions). But once you have finished it, you're ready to take that final step.

User approval to proceed

At the end of the installation stage, whatever may lie ahead, the system is now the user's baby. His approval for you to proceed to the maintenance stage signals his willingness to live with the system as it is. It usually indicates his willingness as well to pay your bill for system development to this point. This approval can be termed the formal transfer to maintenance. It can be labeled the informal presentation of the bill.

So sit back with Colbert and Louis, with all the other teams and users we've discussed, and have a glass of that old dependable French relaxer: champagne. You deserve it!

Among its many functions, Versailles also served as a home—for the Royal Family and about 2,000 others. Things don't really change much, though: the king's wife has the little princess on a "leash" because she won't stand still for the portrait.

Exercises

1. Define the following terms:
   a. Installation.
   b. User training.
   c. Operations training.
   d. Different types of training.
   e. Acceptance testing.
   f. Summary of development.
   g. Turnover to maintenance.
   h. Parallel mode of installation.
   i. Pilot mode of installation.
   j. Immediate mode of installation.
   k. Midinstallation paralysis.

In each of problems 2 through 5, a system and a user organization is given. Discuss how you approach the problem of the users who should be trained to operate the system and to work with the results and inputs to the system.

2. The electric light bulb and electrical systems in the home; home and apartment dwellers.
3. A new truck; owners of trucking firms, drivers.
4. New home; homeowner.
5. New tax laws; general public.

In each of problems 6 through 10, a system and a user organization is given. Suppose you were in charge of training the users. Discuss how you would go about training them.

6. The electric light bulb and electrical systems in the home; home and apartment dwellers.
7. A new truck; truck drivers.
8. New home; homeowner.
9. New tax laws; general public.
10. New payroll system to replace a manual system; payroll department staff.

In each of the situations in problems 11 through 15, you are in charge of conducting acceptance tests. Discuss how you would test the system. Where would you collect data on how the tests went?

11. A new home.
12. Toy gliders made out of balsa wood.

13. An automobile.
14. New tax laws.
15. A fifty-five-mile-per-hour speed limit to replace a sixty-five-mile-per-hour limit.

In each of problems 16 through 20, discuss which type of installation (parallel, pilot, or immediate) you would employ to install the new system.

16. A law to decriminalize the use of a drug.
17. The marketing of a new type of car.
18. A new method of handling your checking account.
19. A new procedure to hire employees.
20. A girlfriend, boyfriend, or lover.

In each of the cases in problems 21 through 25, the system failed after it became operational. Discuss what could have been done during the installation stage to prevent failure.

21. A new accounts payable system was installed. The staff complained of all the extra work needed to use the system. The vendors who were paid from the system complained of late and erroneous payment.
22. A house in a housing development is discovered to have a cracked foundation five months after it was built.
23. Prohibition in the 1920s in the United States.
24. The Munich agreement in 1938.
25. The failure of a system during operation can be traced to the failure of transforming the group I list to the group II list. For each of problems 21 through 25, discuss what group I list elements were not transformed. How could the transformation be improved during the installation stage?

In each of problems 26 through 27, discuss how midinstallation paralysis could occur and what could have been done to prevent it from occurring.

26. Mass rapid transit system in parallel to freeways and buses.
27. Regional shopping centers in suburbs to replace downtown shopping centers.

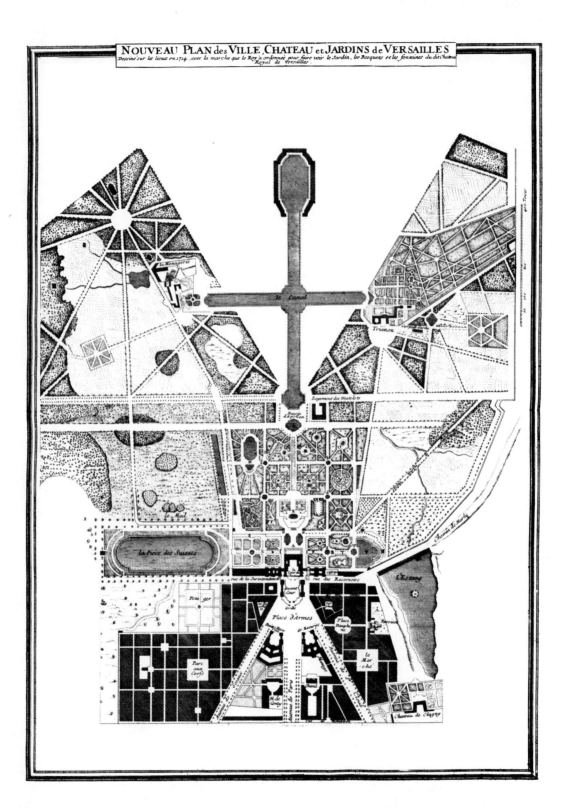

NOUVEAU PLAN des VILLE, CHATEAU et JARDINS de VERSAILLES

Dessiné sur les lieux en 1714, avec la marche que le Roy a ordonné pour faire voir le Jardin, les Bosquets et les fontaines du dit Chatau Royal de Versailles.

9

The Operation Stage: Maintenance

Before we begin our discussion of maintenance, we're going to jump ahead. We're going to discuss both maintenance and enhancement, since they occur simultaneously. The importance of distinguishing between the two is enormous. Maintenance activities may be defined as those activities that are necessary to keep the system operating in the way it was designed and built to operate. Enhancement activities may be defined as those activities that are necessary to help the system operate under new or changing environmental conditions. Maintenance and enhancement are vital to keep a system functioning. The system must not only satisfy existing requirements, but also new ones that arise in the environment. These activities—maintenance and enhancement—consume over fifty percent of the life cycle cost in many systems.

In other words, when you find that the system is beginning to fail in its operations, you must first determine whether the cause of that failure is a new environmental pair which has crept unnoticed into your group I list or something that is simply wearing out from normal use. If you find that a new environmental pair has been added, you face a system enhancement. If it's simply a matter of light bulbs burning out, paint chipping, or plumbing getting clogged, you have a maintenance problem.

195

Making the Environment Work for You

As you're by now no doubt aware, the thrust of system development is to create a way of coping with an environment. We started out by making an exhaustive group I list, which is simply a statement (or series of statements) about the problematic environment. Then we proceeded to try to convert the problematic environment into a sustaining environment by building a system geared to do just that. We transformed the force elements of the group I environmental pairs into force elements of the group II list by adding the system. We've made the environment work for us.

After the system is in operation, the environment should continue to work for the user. When it doesn't, you (the system developer) must embark on maintenance or enhancement activities. These activities are ongoing throughout the useful life of the system and therefore accrue a large amount of the system life cycle costs. During the lifetime of a residential dwelling, for example, more money is put into maintenance than was ever put into building. Such items as light bulbs, toilet paper, brooms, furniture, floor polish, plumbing corrections, and other fix-it activities account for as much as 70 percent of the total funds expended on the house. Check a normal family's budget if you doubt this. After all, with a little bit of luck, the time will come when the mortgage will be paid—but the maintenance of the structure goes

The fabulous Versailles fountains. They still work!

on and on. Versailles, for example, has been maintained more or less continuously now for over three hundred years.

Maintenance is an important and costly activity, but it must be faithfully carried out if the system development activity is to be important and successful. When you find that the system isn't working at peak efficiency, that the environment is regressing back to a problematic state, get out your tools and go to work.

How Do You Begin Maintenance?

Obviously, certain types of maintenance will be dictated by day-to-day operations. A computer facility will have to keep the filters clean on the air-conditioning units or the dust will begin to affect the operation of the equipment. Occupants in a home will have to replace light bulbs and to vacuum rugs.

Other, more serious types of maintenance will be dictated by temporary system failures. Examples of this type of maintenance include replacement of worn-out engine parts in an automobile (your fan belt breaks and your car overheats, so you replace the fan belt).

Basically, there are two types of maintenance:

1. Preventive maintenance, which includes housekeeping activities, periodic inspections, ongoing checking activities (such as an auto mechanic performs when you send your car into the shop for a checkup).

2. Repair maintenance, which includes replacement of nonfunctioning components and periodic updating of existing system (addition to computerized facilities of new data bases as they become available, for example).

The difference between the two types of maintenance is simple: preventive maintenance deals with changes designed to keep a functioning system functioning; repair maintenance deals with changes designed to reinstate a system which has temporarily failed (or will fail shortly). We have included the periodic updating type of maintenance in the repair category because it affects the up-to-date functioning of the system: it restores to a system which would otherwise be outdated its original currency. Figure 9.1 lists some preventive and repair maintenance activities for the Versailles case study, as well as for two modern systems.

FIGURE 9.1

Some Examples of Preventive and Repair Maintenance Activities

	VERSAILLES			CITY STREET MAINTE-NANCE	UNEMPLOYMENT ASSISTANCE
	Physical	**Political**	**Economic**		
PREVENTIVE MAINTENANCE	1. Oil door hinges 2. Check appearance of silver	1. Check on status of specific nobles 2. Update guidelines on style	1. Purchase more French goods routinely 2. Inspect furniture to ensure all is French made	1. Sweep streets 2. Inspect sidewalks	1. Spot-check for meeting eligibility requirements 2. Add staff to prevent possible excessive delays in processing
REPAIR MAINTENANCE	1. Fix broken chairs 2. Repaint weathered walls	1. Reassign some nobles to lesser quarters according to rank 2. Fix procedures to keep two ranks of nobility apart	1. Order furniture from a wider range of French sources 2. Start projects to ensure money is recycled	1. Replace burned-out lights 2. Fill in holes in the streets	1. Close loophole in procedures that adds ineligible people 2. Fix procedures to ensure that checks get produced on time

There is, however, another key to the start of maintenance. It's called a postinstallation review.

The Postinstallation Review

The last formal, scheduled activity of most system development teams is a document called a postinstallation review. It's a task no one wants. Whoever gets stuck with it usually feels like a martyr and attacks it with all the gusto and precision of an army mess hall chef.

The fact that a postinstallation review is not a coveted activity in a system development team makes for a considerable amount of trouble in the real

Sydney's picturesque Opera House fulfills one user requirement perfectly: it is a postcard identification of Australia. The postinstallation review yielded an outstanding flaw, though: shallow, spotty acoustics.

world, however. Why? Because these reviews are frequently shoddy, incomplete, surface-skimming pretenses which serve no real purpose. Why does this kind of review exist at all? It serves (or should serve) a real and important purpose. It may be yesterday's potatoes to someone who would prefer to work on a new project, but it's the *only* backward glance that you have at a system that has been completed and is in operation. It's your last chance to prevent system failure and to ensure system success. It's both an invaluable analytical tool for the system development team (What did you leave out? What did you fail to do that should have been done?) and an essential warning flag for the user. When properly performed, the postinstallation review is a careful and precise measuring and evaluation activity.

Colbert's Postinstallation Review

Let's say that Louis has moved into Versailles with his court, that the move was completed six months ago, and that everyone has been happily installed at the new château. At least they are happy as far as you can tell on the surface. It now becomes incumbent on Colbert to complete a postinstallation review.

What criteria should he use in compiling his review? He must go back to his files and retrieve the user requirements his team compiled during the analysis stage. Sounds like ancient history, doesn't it? The analysis stage happened so long ago, you may have totally lost sight of what the user requirements really were. If you have, refer to figure 5.4, page 112, to refresh your memory.

Why use an old document like the user requirements? Why not use the system design documents? The user requirements are still the user's latest statement on what he wants and needs; they are still the bible of the system. If the system doesn't conform to the user requirements, it doesn't function properly.

We listed six criteria in the user requirements. (In reality there would have been more, and they would have been more detailed. But, since we are in a realm of inventive history, we'll work with those six requirements here.) Figure 9.2 gives some idea of the results that Colbert finds during his probe for information. As you can see, he encounters more than just maintenance information.

FIGURE 9.2

Data Uncovered During Postinstallation Review at Versailles

USER REQUIREMENT

1. Segregate members of government and nobility; members of government get less comfortable quarters than nobility

2. Make Versailles a fun place to enjoy, not work

3. Build protocol into palace

4. Decorate expensively to require expensive living by inhabitants

5. Install facilities for court entertainments

6. Subsidize noble big spenders

POSTINSTALLATION REVIEW FINDINGS

1. Unequal quarters for equal rank found
2. A commoner-minister is located too near the king's quarters
3. A duke is mistakenly housed among commoners

4. Zoo animals are not well maintained; some appear seedy
5. More exotic animals are needed
6. Not a broad enough range of entertainments

7. Court life has drifted somewhat away from Hall of Mirrors
8. The shortage of chairs temporarily has some higher nobles eating with lesser nobles

9. Not enough mirrors
10. A cracked mirror is found in a room on the first floor
11. Several blank walls have been found; they need decoration

12. Opera house acoustics are imperfect
13. Some of the fountains need to be cleaned
14. An additional fountain is needed in a small garden
15. There is a danger of fire from the fireworks stand

16. The heraldic college is found to be too susceptible to bribes
17. Some of the personal gifts from the king are of less than adequate quality

Every postinstallation review unearths a variety of information. Most of it falls into one of three categories:

1. Maintenance information.

2. Enhancement requests.

3. Hindsight information on the system development process.

Before the review is written, these categories must be sifted and sorted out. Only the information that falls in categories one or two belongs in the review.

A Word About Witch-Hunts

There is a strong temptation to embark on an accusatory expedition during a postinstallation review. With the perfect vision that hindsight usually brings, the goofs that were made during the development process frequently stand out with unusual clarity. It's always a time of trial for the A&D member who conducts the review.

Let's say, for example, that Colbert's staff finds, while researching a postinstallation review, that a certain architect has consistently specified staircases that are too steep for court ladies to feel comfortable climbing. Wouldn't it be easy for Colbert to descend like the wrath of an angry Valkyrie on that architect? Wouldn't it be natural? After all, those staircases may be a serious blight on Colbert's master achievement in building the palace.

But what can Colbert achieve by chewing out the architect? The ladies of the court will still have trouble going up the staircases, whether the architect's head rolls or not. And, if the staircases are the only problem, then the architect did a lot (a lot!) of things right. Colbert may well find himself working in collaboration with the architect again someday. The answer is that there's very little to be achieved by reprimands at this stage. The emphasis and thrust of the review must be to correct whatever needs correcting, to enhance whatever needs enhancement, and not to find fault with what's now in the past.

The point we're trying to make here is that so-called witch-hunts are always counterproductive. The mistakes of the past can be valuable in the future; they're valuable learning devices. If Colbert notifies the architect in question that the staircases are too steep, that architect probably won't make the same mistake again. He may even be motivated to find out why he designed such staircases. If he's accused of fouling up the system, he has no recourse but to protect himself, which may involve a lengthy chain of recriminations and accusations.

Standard Review Periods

Many modern systems have such reviews built into them. We call them warranties. When you buy a new car, you get with it a warranty for a certain period of time, usually a year. Prior to the end of that warranty, you should have the system development team conduct a postinstallation review of the car as a system. That postinstallation review usually consists of a complete, exhaustive checkup on how the car is running.

There's a feeling in modern business that warranties exist even when they aren't stated, and that feeling extends itself particularly to system development teams. Anyone who hires a team to develop a system should be entitled to a final study after a period of operation. Most users will demand it. System development teams ought to demand it even if the users don't—if for no other reason than just to close out the books. From the time that problems are remedied in postinstallation, any new problems that occur are the property of the user (such as the water pump that fails in your new car after 20,000 miles).

The Mechanics of the Review

You have the user requirements in hand. Your first step is to conduct interviews with whatever key user personnel you think may have reliable information about how the system is functioning. These interviews should have as a basic plan the following three questions:

1. Does the system meet the needs originally specified by the user and his operations team?

2. Is the user organization getting the benefits they thought they would get from the system?

3. What new needs does the user have for the system?

Nobody likes to be quizzed or questioned about his work, and you may be viewed as an intruder during these interviews. Be as polite and inoffensive as you can be, but make sure you gather information that's accurate and complete. Make certain that the people you interview are aware that you aren't checking up on them, but that you're only trying to see how well the system is working.

These interviews should point you in some directions for data collection. Remember to collect data on a neutral subject; don't look for backing for a particular idea you may have, because you can find substantiating evidence for almost anything if you look hard enough for it. Collect your data impartially. Remember that what you're looking to do is to find out whether the system lives up to your expectations and the expectations of the user organization.

Once you have the necessary data in hand for your analysis, you must proceed with the measurement of the system's success. To do that, your first task must be to decide what criteria to use and what yardsticks to apply. Sound familiar? You may find it helpful at this point to refer back to chapter 4, in which we established feasibility criteria and yardsticks. (Refer also to chapter 7 for measurement during testing.) This is a similar task, with an important difference.

When you worked up feasibility criteria, you were working with castles in the air. Nothing was there except what you decided *should* be there. There was a good side to that and a difficult side. In the plus column, there was the fact that you didn't have to deal with nitty-gritty realities (which makes everything appear a bit neater than sweaty reality sometimes is). In the minus column, there was the reality that everything you theorized was unreal: theory. You had nothing to stake your statements on except your unfailingly good judgment.

Now you're in a different pickle, but you're still sandwiched in by a good and bad squeeze play. We'll give you the good side first to soften the blow. The advantage now is that you have a real, functioning system to measure. There's very little guesswork about the work you're doing now; it's real measurement and evaluation of a real system. The fact that you're no longer judging projections and plans, but realities, is a very strong plus. In addition to that, you should have the user available to help you decide how well things are running overall.

Then there's the sticker: reality is just not neat. If there's one sure thing you can say about a functioning system, it's this: it doesn't function equally throughout. What does that mean? It means that you're now working with people and machines, or buildings, or whatever, and there are few generalizations that will apply throughout such a flesh-and-bricks reality. In addition, the data will be harder to obtain because you now have to get it from people who are busy doing something besides what you want them to do for you.

Say, for instance, you're doing a postinstallation review of a new metal ore processing plant that has been your pride and joy. You decide that, in order to complete your study, you need some specialized data. You can get that data most easily by finding out just how long it takes a load of ore to get from one end of the plant to the other (through all the processes it must experience before becoming ingots). So you develop a series of special time cards to tell you just how long each function is typically taking.

You walk your time cards into the plant manager's office and ask if he would mind having his people use them for a couple of days. He looks at you as if you had three eyes. Why? What you're asking him to do:

1. Violates his union contract by requiring his men to perform clerical duties.

2. Slashes away at his efficiency by adding work to an already busy schedule.

3. Threatens him, because he thinks you're out to get him.

4. Angers him, because you didn't think of all that yourself.

If you're lucky, he'll refuse your request. If you're not so lucky, he may comply. If he does comply, you're liable to get the sloppiest, least reliable data that ever hit your hot notebook. He's going to hand those time cards over to the supervisory staff and say something like, "That high-powered what's-his-name from the systems group wants you guys to fill these out. I don't care what you do with them." That attitude will pervade your data and results.

How Do You Get Dependable Data?

Getting dependable data isn't as hard as you might think, if you follow a couple of simple rules:

1. Use what's available first. Analyze whatever statistics are generated in the regular course of business before you go around disturbing people at work. You should be able to get your hands on production statistics, personnel statistics (number of staff hours devoted to each department, for example), and cost statistics—all rather easily.

2. Don't come on like a forms-happy efficiency expert when you talk to people.

One of the surest ways to alienate people in the user organization is to ask them to fill out a form. Everybody hates forms; they're one of the banes of existence in our society. You can only generate dislike by using them. If you must collect your data on forms, then put it on forms yourself from notes you have made in personal interviews.

A word of warning: you may not have all the time you want in this study. You frequently get only a limited amount of time with each member of the user organization because they're all quite busy coping with the system you foisted off on them. Make your interviews count, and make them as short as they can possibly be. If you do that, you should come out all right.

I Never Saw This Before

Here we arrive at something else you should be on the lookout for: user innovations. Now that the user owns the system, he can do anything he wants with it (as long as it's legal). You'll find out that he frequently does just that: anything he wants.

Consider this case study: A community college district had a complete set of blueprints for all the new buildings in the new area of campus. They had the blueprints because you gave them to them. Somebody from your group went out to write a postinstallation review on the way the buildings were functioning. He found out the hard way that the user had made some changes. He was nearly electrocuted making the discovery that some of the wiring had been changed, but the changes were not reflected on the original blueprints.

Although death-dealing innovations are rare, almost every user begins altering the system from the ribbon-cutting ceremony onward. Be aware of user innovations when you are analyzing data. Some of the user innovations may be sensible, functional changes. Others may be the equivalent of chewing gum and chicken wire. Part of your duty in this review is to decide which alterations are viable and which may cause system failure.

Maintenance Versus Enhancement

This is your first stab at enhancements to the system. Virtually every system of even normal complexity will begin to demand enhancements almost immediately. Normal operations will unearth the damndest things—things nobody could have foreseen. They may be easy to take care of; they may not. And they may require enhancements. Part of your job during the postinstallation review is to note requests for change and pass them back to your system development team for action.

Finishing the Review

The review has to be a living document that deals totally with two main issues:

1. How well is the system operating in relation to what the user originally wanted it to do?

2. What steps must be taken to maintain or enhance operation?

In addition, it must be a study that is complete enough to be a tool in analyzing the performance of your system development team during the previous stages of the life cycle. Fear not that it will be used as a tool of internal analysis. There is an obvious comparison between the postinstallation review and the project summary. Together, the two documents should provide a learning tool for the members of your group. It is a good idea to keep that function in mind as you write.

So you've got it done. Now what?

Maintenance Decision Making

We discussed earlier (chapter 2, page 36) the three levels of management decision making: strategic, managerial, and tactical. All three of those levels must again come into play during the operation stage. All three levels have exceedingly important roles to play now.

Tactical Decision Making

Maintenance makes varied demands of tactical decision makers. Most typically, they are demands of this sort:

1. Fix a malfunctioning component.

2. Paint, sweep, clean, or replace parts.

3. Answer a procedural question.

4. Clarify a technical question.

5. Modify training materials that are not functioning as well as anticipated.

As before, these are all A&D functions. There are as many others as there are types of systems, but they share a common attribute: they're immediate demands that require no substantial change in the design of the system. They're all essentially preventive maintenance, and are frequently performed by the user organization.

Managerial Decision Making

Things start here to get a bit heavier, more complex. There are basically two types of managerial decision making in this stage:

1. Fix a system which has ceased to function.

2. Forecast and provide for future tactical maintenance.

What do these two types of decisions have in common? They both have a discernible impact on the system itself, and they both require trade-offs that involve outside contact.

The budgeting of tactical maintenance is a managerial maintenance task. It's the managerial decision maker who must see to it that there are enough light bulbs to take care of normal use and that a fair price is paid for them. It's the managerial decision maker who must provide staff for tactical maintenance as well. If there's no fix-it man around, whose fault is it?

Managerial decison making is also frequently performed by the user organization, although the user is entitled to ask you to help as well. A frequent managerial solution to mechanical maintenance problems is a service contract—an agreement which allows budgetary and time planning regularity in the user organization.

Strategic Decision Making

What's left? The obvious strategic decision is an ongoing one: is the system functioning well enough to continue its life beyond this moment? In other words, shall we kill it?

The strategic decision maker must keep his eye constantly fixed on the relationship between the group I and group II environments that make up the system. When there's a malfunction or a stoppage, the strategic decision maker must find out whether the group II solution is still a valid solution to the problem. He must also determine whether the group I list as compiled during the earlier stages is still an adequate and accurate summation of the problematic environment.

The Importance of the Group I and Group II Lists

Bet you thought we left those old lists behind long ago. Well, remember they're still the keys that unlock the whole system. If your postinstallation review validates the group II list (confirms that the system actually did manage to create a sustaining environment equivalent to the one you projected), your system development process was successfully carried out. If your review invalidates the group II list (finds that the group II list and the system as it operates do not jibe), your system development process was a failure to the extent that the two don't match.

If you find that the group I and group II lists are still valid and applicable to the situation the user finds himself in, and that the group II list is validated, then the system is functioning and needs nothing more than maintenance. If either the group I or group II lists changes, you're involved in a strategic decision that can be phrased as follows: Do we kill it or enhance it?

We'll be getting into the extreme importance of the group I and group II lists in the next chapter. Don't throw them away; you'll be needing them again.

The Importance of the Maintenance Stage

During a normal life cycle, maintenance and enhancement costs can account for up to 70 percent of all cost. With the passage of time, normal life of parts, salaries of personnel, repairs, and more, many systems spend more money existing than they ever did just arriving at maturity. It would be foolishness bordering on idiocy to abandon the structured, conscious approach that defines the system development process just when big money is going to be spent.

The log in figure 9.3, presented in two versions, shows just how important a systematic approach to maintenance can be.

FIGURE 9.3

Sample Maintenance Logs for a Housing Project

LOG 1

1. Fixed water leak on third floor; plaster damage.

2. Inspector came by and found rats on the fifth floor (must be in other parts of building).

3. Paid $28 for cutting grass to John.

4. Tenant in 3C complained about tenant in 2C.

5. Tenant in 4G skipped out without paying for utilities.

LOG 2

Water leak in kitchen of 3F (drain plugged up). Repaired minor plaster damage to ceiling of 2F.

Inspector Clive Smith (Badge Number 654) examined the building for rodents. Trails of two rats were found. Traps were set.

Paid $28 to John Hoyle for cutting grass (check number 5639).

H. Jones in 3C complained about excessive noise after 10 P.M. coming from 2C (Mary Smith's apartment). Requested quiet. No problems later.

Naomi Tollquist left apartment 4G without paying utility bill of $75. Filed small claims court action.

Maintenance at Versailles

It's been going on for three hundred years now. Colbert died in 1683; Louis passed to his reward in 1715. And Versailles is still going strong. It has been used as a palace, as a revolutionary seat of government, as a hospital, as an international negotiations hall, and as a museum.

And on a sunny spring day, you can still walk through those wonderful gardens, see the fountains that Louis built, and smell the flowers that Louis pianted. It didn't stay that way by itself.

The manicured gardens of Versailles survive Louis's kingship—and the entire ancien regime *by 200 years. The key is careful maintenance (something Louis XIV's successors have never been accused of in relation to the monarchy).*

Exercises

1. Define the following terms:
   a. Maintenance.
   b. Enhancement.
   c. Preventive maintenance.
   d. Environmental factor maintenance.
   e. Maintenance requests.
   f. Postimplementation review.
   g. Measurement.

In each of problems 2 through 6, identify which of the changes are maintenance and which are enhancements. Give reasons for your answers.

2. An automobile: fixing a tire; adding a smog device to an old car; putting in air conditioning.
3. A house: adding a room; painting a room; fixing plumbing; replacing galvanized pipe with copper pipe.
4. A payroll system: automating the system; changing input forms to the system; training new personnel.
5. Law enforcement system: adding more patrol cars; imposing harsher prison sentences; training police due to new court decisions.
6. Energy exploration and production: shipping more natural gas across state boundaries; drilling more wells due to price increases; antitrust actions.
7. For problems 2 through 6, identify which changes are due to preventive maintenance and which are due to environmental factors.

It has been said that maintenance work can introduce errors more serious than the errors being fixed in maintenance. In each of the cases in problems 8 through 12, propose four errors which could occur due to faulty maintenance.

8. A house: fixing an electrical short and outlet plug.
9. An automobile: doing a tune-up.
10. The tax structure: reducing the tax loopholes and shelters.
11. The criminal justice system: imposing mandatory jail sentences.
12. A bus system: reducing the number of lines in one area.

This chapter addressed the postimplementation review. In each of problems 13 through 18, you are to identify data sources for conducting the review.

13. A new tax law to increase federal revenues from gasoline use.
14. The Bay Area Rapid Transit System (BART) in the San Francisco area.
15. A new electric car.
16. A new fertilizer plant.
17. A new Air Force missile.
18. Discuss why preventive maintenance is cheaper and easier to handle than environmental factor maintenance.

In each of the cases in problems 19 through 23, a system is given along with some of the environmental factors that may interact with the system. For each problem, propose five potential changes that could come from each environmental factor.

19. An automobile: the federal government; the state government; an oil company.
20. The telephone system: technology companies; the federal government.
21. You: the federal government; the company you might work for; your boyfriends or girlfriends.
22. A county dump: the Environmental Protection Agency; the manufacturers of bulldozers and earth-moving equipment; users of the dump; developers of houses.
23. The tobacco industry: Department of Health, Education, and Welfare; consumers; environmental groups.

In each of problems 24 and 25, a system is given together with a list of maintenance and enhancement change requests. Discuss which of the requests would be grouped together.

24. A house:
   a. A roof leak over a bedroom.
   b. Insulation to reduce energy needs.
   c. A lack of heat in one room (leak in duct suspected).
   d. A new sprinkler head.
   e. Drain the hot water heater.

    f. Install a new light switch for a room.
    g. Paint a wall.
25. A downtown urban area:
    a. Increase the number of police patrols.
    b. Establish a neighborhood health emergency clinic.
    c. Add a new bus line.
    d. Demolition of four old buildings.
    e. Plant trees and shrubs along the streets.
    f. Establish a lighting district to improve night lighting.

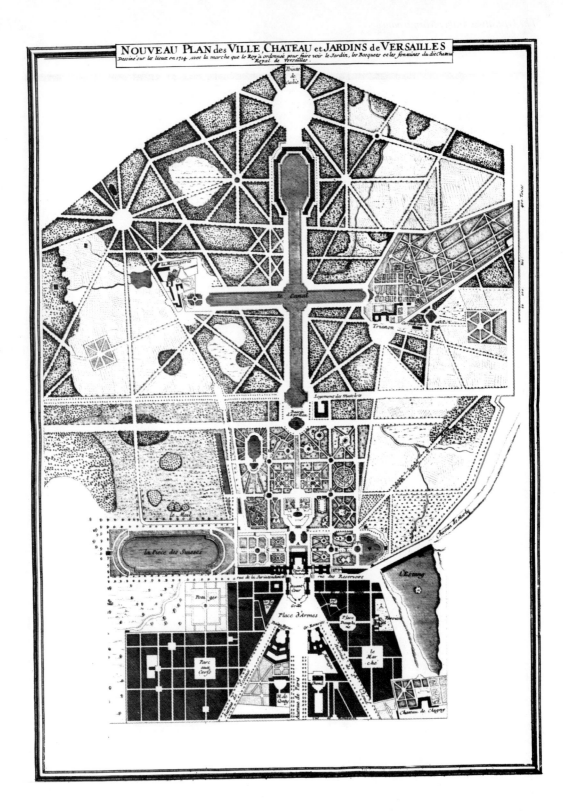

NOUVEAU PLAN des VILLE, CHATEAU et JARDINS de VERSAILLES

10

The Operation Stage: Enhancement

When Louis XIV died in 1715, Versailles was abandoned overnight. In the period of one day, the court reestablished itself at the drafty, uncomfortable Louvre in Paris (which had been uninhabited for nearly fifty years by then). Why? They ran from Versailles because the great château represented a system that was failing. After seventy-two years of rule, Louis died a spent man, king of a bankrupt country, figurehead of a warring royal family and a restless nobility. The two solutions that Versailles was supposed to implement seemed to have failed. And so the château was abandoned, just as Louis's ideas were abandoned.

But it's well-known that Versailles was inhabited again. Louis XVI received Benjamin Franklin at Versailles. Marie Antoinette and Louis XVI were arrested there and hauled back to Paris by the mob of the Revolution. Napoleon and Josephine moved in for a short time. Louis XVIII, Charles X, and Louis Philippe all spent time there. Napoleon III and Eugenie entertained Queen Victoria there in 1856.

What happened to make people move back in? You've got it: what happened was enhancement. The system represented by the château was altered to fit in more comfortably with altered circumstances. It's something that happens to most successful systems, although it's generally ignored by textbooks and by system development groups.

An example of maintenance and enhancement. The graceful Petit Trianon at Versailles, once the private escape of Marie Antoinette, is now a charming—and still beautiful—museum on the château grounds.

Catch as Catch Can

The rule in the real world where enhancements are in question seems to be, "Put a patch on it." Seldom do even the most careful and systematic system development teams approach enhancement as enhancement. It is split up and treated as either *(a)* another aspect of maintenance or *(b)* a new system. Both of those treatments are wrong—no question about it, wrong. But it's frequently easier to lump things under large headings when they may cause trouble just sitting there.

Failure

You drive your car out of the shop; you've had an air-conditioning unit installed, and you're feeling satisfied with yourself. Then you find that your car won't climb the hill to your house easily anymore because the air conditioning draws too much power off the engine. Or you switch on your new color television and all the circuit breakers in the house turn off. At the office, you request another report from the computer center and you find that the new request delays your regular reports an extra three weeks because of work overload.

A city accedes to union demands after a long sanitation strike and teeters on the edge of bankruptcy because of the new problems meeting those demands makes in the budget. A respected university finds its degrees are increasingly devalued because of its policy of passing through students who do not meet normal degree requirements. The nation outlaws offshore oil drilling for valid environmental reasons and then finds itself immersed in a depression because imported oil increases dramatically in price.

This list could go on for days, but we'll stop here. What do all the situations listed above have in common? Failure, for one thing. For another, they share a sloppy approach to system enhancement, which in each case has induced failure. What's more, each of the enhancements has been in the normal course of business; nothing unreasonable has been done. Nothing has been drastically altered (to all intents and purposes), but now the system is dying. Enhancements are absolutely critical in the life cycle of most systems. The only systems that escape this stage are disposable ones: you use up a ballpoint pen and throw it away—simple death.

An ad hoc approach to enhancement is a sure path to suicide, make no mistake about it. You can't cure a disease with a bandage, and you can't cure an ailing system with an add-on. Anytime you add something to a system, you make a trade-off somewhere else. With the car mentioned above, you lost acceleration. With the color television, you overloaded the electrical circuitry. And so on.

Types of Enhancement

Basically, there are two types of enhancement to any given system:

1. Enhancements requested by the user organization on the basis of functionality. An example of this type of enhancement is a rapid transit system that asks for bus-devoted freeway lanes to facilitate rush hour bus traffic. The bus passengers have increased over a period of time to a population that justifies such a request.

2. Enhancements dictated by a change in the environment. The new disclosure laws pertaining to bank loans are a good example here. Banks are now required to state interest in simple per-year numbers on loan applications. Where before your bank charged 1½ percent per month, you now are told that it charges 18 percent per year. The change in law represents an environmental demand upon the bank loan system.

Enhancements may be defined as changes in a system caused by alterations in the group I and/or group II lists. How does this definition apply to the two types of enhancement mentioned above?

User Enhancements

When a user organization requests a system enhancement that is designed to improve operation, he's altering the old group II list. He's saying, in effect, that the response of the sustaining environment to the problematic environment must be improved. Let's take as an example the rapid transit request mentioned in the section above. The group I/group II lists for the system might look in part like this:

GROUP I
1. C = Downtown arteries are clogged during rush hour.
   F = Private one-passenger cars are the primary mode of transportation.

GROUP II
1. C = Institute bus lines on main traffic arteries.
   F = Provide alternative to private one-passenger transportation.

At the time that these lists were formulated, they were accurate perceptions of the two environments. When the bus system was installed, it worked. It decreased the amount of traffic significantly. Unfortunately, the bus transportation didn't match the private car mode in *speed* or *convenience*. The rapid transit district now believes—probably correctly—that more commuters will ride the buses if the buses are given a special freeway lane to get them downtown faster than private cars are able to get downtown.

What has happened? The group II list has been altered. The revised group I/group II lists might appear like this:

GROUP I
1. C = Downtown arteries are clogged during rush hour.

   F = Private one-passenger cars are the primary mode of transportation.

GROUP II
1. C = Institute bus lines on main traffic arteries, *with devoted freeway lanes.*
   F = Provide *fast* alternative to private one-passenger transportation.

The group I list has remained unchanged even though the situation has gotten somewhat better with the introduction of the buses. The response to that list, that problematic environment, has changed. This type of enhancement is referred to as a *user enhancement.*

Environmental Enhancements

A user enhancement is pretty straightforward; it's the response of a system operator with some experience in the system. It may be the brainchild of someone intimately acquainted with the system; it may be an obvious choice of alteration to everyone. At any rate, it's usually a discretionary alteration: the system will continue to function without it but will function better with it.

That isn't the case with environmental enhancements. The system may well cease to function *at all* with them. At the very least, the system will be crippled without them. What are they? *Environmental enhancements* are alterations dictated by a change in the system's group I list. Take the bank loan example. Prior to the passage of the disclosure laws, the group I/group II lists may have looked like this:

GROUP I	GROUP II
1. C = Need to sell revolving credit programs to customers. F = Need to increase profits.	1. C = State interest rates in smallest possible units. F = Real yearly interest rates will scare customers away.

The system functioned well when instituted; the bank sold its credit card system to a good number of its clients. The interest rates were stated as 1½ percent per month, and customers saw that number as little enough to pay for the convenience of their credit cards.

Then the loan disclosure laws were passed (truth-in-lending acts), and the whole situation changed. Now the group I/group II lists look like this:

GROUP I	GROUP II
1. C = Need to sell revolving credit programs to customers. F = Need to increase profits.	1. C = Base advertising and promotion on convenience aspect of card only. F = Convenience is only strong selling point.
2. C = *Must state yearly interest rates on all loans.* F = *New law requirements.*	2. C = *Develop printed formula for stating interest rates.* F = *Customers don't read formula statements on loans.*

The whole nature of this segment of the system has been changed by the introduction of a new group I environmental pair. The functioning of the system as a whole depends on this change. The new truth-in-lending requirements invalidate all the old application forms, all the old monthly statement forms, all the old advertising approaches. The system will cease to function if changes aren't made because the government won't allow things to function as they were.

Nevertheless, the system as originally conceived had many components, and as a whole it will continue to work well if these changes are made. The revolving credit system is a big money-maker for the financial institutions; it's an important convenience for the customers of those financial institutions.

We're becoming a credit society, and the base assumptions behind the program are still valid and vigorous. The system will be retained (not killed) because it's still profitable and because it can be altered to comply with the new regulations. This type of alteration is called an environmental enhancement.

Enhancements at Versailles

During its three-hundred-year history, the château has been modified and enhanced countless times. It was made a center of world intrigue under Louis XVI and Marie Antoinette by becoming a focal point for Austrian infiltration of France and later by becoming a symbol of aid to American revolutionaries. It was used by the Revolution as a symbol of wickedness and as an invaluable source of capital. It has served as hospital, as diplomatic capitol (the World War I armistice was signed there), and as museum. Figure 10.1 discusses a very few of these enhancements and classifies them as either user or environmental enhancements. We have chosen two periods of history for figure 10.1, the Revolution and the de Gaulle ascendancy, for no reasons other than brevity and simplicity. Enhancements have been constant throughout the centuries.

Signing of the Treaty of Versailles, which ended World War I and founded the League of Nations. The great Galerie des Glaces faces the gardens.

FIGURE 10.1

Enhancements at Versailles

LOUIS XIV[1]

REVOLUTION (ONE HUNDRED YEARS LATER)	
GROUP I	GROUP II

LOUIS XIV[1]

C = Flaccid economy
F = Civil disorder

C = Tradition is impeding progress
F = Tradition is embedded in all current practices

C = There are no incentives to industry
F = The government is not spending public works money

C = The nobility is too powerful
F = The power is enhanced and magnified by physical location

GROUP I

C = Boom economy
F = Civil disorder

C = Tradition is impeding progress
F = Tradition is embedded in all current practices

C = No programs
F = The government is not spending public works money

C = The status of the monarchy is too powerful
F = The power is enhanced and magnified by physical location

GROUP II

C = No response
F = No response
Environmental enhancement

C = Thorough changes (e.g., change church, calendar)
F = Erase tradition
User enhancement

C = Sell furniture
F = Need to transfer money
Environmental enhancement

C = Move to Paris
F = Paris is the traditional center of power
User enhancement[2]

---

1. The forces and conditions are drawn from those on page 136, figure 6.9, numbers 6, 9, 10, and 13, respectively.
2. Since the Revolutionary government replaced the monarchy, it causes a simple change in terminology in the group I list—not a substantive change.

Nevertheless, the Revolution and the rise of de Gaulle are two turning points of French history, and both used Versailles as a key to national politics and economics. As the figure points out, the Revolution sought to dismantle

POST–WORLD WAR II MUSEUM
(NEARLY TWO HUNDRED FIFTY YEARS LATER)

GROUP I

C = Flaccid economy
F = Civil disorder

C = Tradition is key to
     progress
F = Versailles is symbol of
     past glory

C = There are no incentives
     to industry
F = The government is not
     spending public works
     money

C = The National Assembly
     is too powerful
F = The power is enhanced
     and magnified by
     physical location

GROUP II

C = Emphasize history
F = History will unify France
     Environmental
     enhancement

C = Restore tradition
F = Redevelop French
     esteem in world
     Environmental
     enhancement

C = Increase employment
F = Need workmen at
     Versailles
     User enhancement

C = Presidential dwelling at
     Rambouillet near
     Versailles (de Gaulle)
F = Want to encourage
     comparison with past
     glory
     User enhancement

the symbol of royal decadence and sell the contents. De Gaulle sought to establish himself as a latter-day Louis by living at Rambouillet, a more modest establishment in the Versailles neighborhood, and he tried (successfully) to reestablish French luxury superiority (and tourism) by reviving interest in the glories of the French past.

Recognizing the Need for Enhancements

One of the largest problems with enhancements is recognizing when they are necessary. User enhancements are no problem; if they aren't perceived, they simply aren't necessary. Environmental enhancements, on the other hand, are a real perception problem. They can come from anywhere because they represent changes in the environment. If your income tax statement is audited and found to be deficient, you may be the victim of a changed group I list of which you were totally unaware. You simply didn't know that the rules had changed. If your product begins to decline in sales, it could be due to any number of environmental changes—and your system development group will have to undertake a study, which could be of major proportions, to determine what's causing the dip. A Mars probe that ceases to operate is working in an unknown environment, and a total analysis is necessary before even a glimmer of an idea can be formulated.

   The procedures necessary to recognize group I alterations vary considerably from industry to industry and from system to system. One thing that is common, however, is the tendency to confuse enhancements with maintenance. The key to that problem is the group I and group II lists. If those lists aren't affected by the need for system alteration, then you're faced with a maintenance problem. If the group II list is affected but not the group I, you have a user enhancement. If the group I list changes, brace yourself, because you're facing an environmental enhancement.

   The alternative to an environmental enhancement is a decision to kill the system.

How to Implement an Enhancement

Essentially, in order to effect an enhancement, you must trace through the steps of the entire life cycle in miniature. The first step is a brief look at feasibility, then at analysis, design, construction, and implementation. There is, however, an important difference.

Every enhancement has a built-in group I environmental pair generated by the system. That group I problematic pair is the group II solution that is currently malfunctioning. If that sounds a bit confusing, let's have another look at the rapid transit system we discussed earlier (pages 218–219).

Notice that the group I list didn't change for the entire system. However, we're going to work with a phantom group I environmental pair. The phantom pair is the segment of the group II list that we want to improve. Why? Because now that the system is in operation, every bit of the system is a piece of the environment *as it exists now.* If we seek to change anything, we're altering the environment.

So, we perform a miniature feasibility study on the group II pair that should be improved and we survey the group I list for domino principle changes. The key question here is: *Will the group I list remain the same if I change the operating environment?* In the case of the rapid transit system, it's a very germane question. As the city of Los Angeles learned in 1976, the idea of a bus-devoted lane was unworkable because of resistance from passenger-car drivers. The group I list, which involved public support for rapid transit, changed as a result of a simple change in the group II list. It was found, at last, that in order to operate a bus lane effectively, the lane had to be physically separate from the rest of the freeway. That solution was adopted in a single instance, and has worked satisfactorily. But, because the Los Angeles RTD tried a group II change without an adequate survey of the group I repercussions, the entire concept failed and was discarded.

Once you have completed a feasibility study of the proposed enhancement detailing all alternatives and their trade-offs, you proceed through the rest of the standard life cycle *as you would with a new system.* This brings us to the next obvious question: should we, instead of enhancing the system, kill it?

Enhancement Versus Death

The question of enhancement versus death is the central part of the enhancement process. Its ramifications are probably too large to generalize upon, but some guidelines may help. Figure 10.2 lists a series of questions to help in the decision.

FIGURE 10.2

Questions to Ask in Considering Enhancement versus Death

1. Is enhancement likely to keep the system running for a long period without repeated enhancement?

2. Is the new setting for the system so different from that of the original group I list that building a new system is cheaper?

3. Has the old system been so compromised that no one has any respect for the system?

4. If enhancements are made, is it likely that they can accommodate future changes?

5. Has the user organization changed so that there's no comfortable relationship with the system?

6. Has technological change mandated that a new system be built that depends on the new technology?

7. Does the list of enhancements indicate that a great deal of the system will have to be changed?

8. Have past enhancements and maintenance made the present system so unworkable that change without rebuilding is out of the question?

9. Has the documentation and knowledge of the system so deteriorated that no one really knows how the system works and what is correct output from the system?

10. Is the organization falling behind its competition because the competition has adopted new systems?

11. Have the criteria for measuring system performance changed so much that the system appears bad under the new criteria?

12. Are there side effects from the use of the system that seriously impair its usefulness?

13. Is there a need for a supersystem—one that will encompass the entire existing system as a subsystem?

14. Have economic conditions changed so much that it's impractical to continue to operate the system?

15. Has the focus of the organization changed so much that the system is now irrelevant?

Generally speaking, if you're contemplating a user enhancement that won't affect the group I list appreciably, you're safe in assuming that you're better off to enhance than to kill. Other situations aren't so clear-cut. Every environmental enhancement carries with it a difficult decision about whether to kill the existing system and start over.

Obviously, if you plan to alter the group I/II list, you're planning to kill at least part of the system. The fact that you're viewing part of your group II list (which was intended to sustain the solution) as a segment of the problem means that you have already decided on an amputation of part of the system solution that you labored so hard for just a short while ago. Where does that leave you? With a cup of strong coffee on the table and a question mark in the air.

Failure Versus Death

As you probably recall, we made a distinction between these two concepts in chapter 1. Death is not a form of failure; it's simply an end to the life cycle. If your ballpoint pen runs out of ink, the system hasn't failed; it has died. If your ballpoint pen runs out of ink while you are trying to take notes at a lecture, your system has failed *and* died. But the second system is more complex than the first, and it includes your decision to take only one pen to the lecture. If you had taken two pens, there would have been no problem.

At some point, every system will die. If a system survives beyond reasonable expectation, it probably has experienced a number of enhancements that have effectively converted it to a new system over a period of time. Such is the case with the palace of Versailles, which has now become a museum, rather than a dwelling. Such is the case with the United States Constitution, which, through amendments and interpretations, represents a much different concept than it did originally. The same is true of other systems noted for their survival: the English monarchy, the wheel, and the use of fire to cook food.

In cases like these, where failure never occurred, the systems were always in basic tune with the group I list—at least in close enough synchronization to roll with the punches. Very few systems have this instinct for change, and very few systems need it.

Tourists crowd around the enormous equestrian statue of Louis XIV which dominates the entrance to the château. No longer a home of kings, Versailles is still the dream of millions of visitors.

An Honorable and Natural Death

The death of a system usually heralds the birth of a new system, and it's usually viewed as a birth, rather than a death. Consider the following two cases:

A. Payroll System, Manual

The Tricorps payroll department has been operating a manual payroll system for the last forty-five years. It has never failed; it has never missed a payday. Nevertheless, with the advent of new government regulations and the need for more extensive record keeping, a decision was made at Tricorps to implement a new computerized payroll system.

After a careful and successful developmental life cycle, the new system is implemented and the old system is phased out. Death. Nothing in it to resemble failure, because until the day its doors were closed, the old payroll department never made a serious error. It just had no real capacity for expansion, and it was in the long run more costly in terms of legal expenses, records maintenance, and so on.

B. An Election-Year Defeat

Senator Jones has been the incumbent junior senator from your state for the past six years. He stands for reelection, and fights a younger man with an outstanding record as a state politician. Senator Jones loses the election. Death, for Senator Jones's system in Washington. And, while there may be some consternation in the papers initially, the incident is viewed more as the birth of a new senator than the death of an old one.

Nevertheless, Senator Jones's defeat is a mixture of failure and death. His system has failed because it did not secure his reelection. But the electoral system—and the people of his electorate—haven't failed; they have simply changed representatives in the Senate. From the viewpoint of the federal system, there's no failure in Senator Jones's defeat.

Just as the death of a system ushers in the birth of a new system, failure of a system ushers in a period of chaos. Consider these two examples:

A. A Senior Executive Dies

On his way to a high-level audit meeting, a senior vice-president of a large corporation dies in a plane crash. Subsequent to his death, the corporation finds to its dismay that he was the only person with access to a large body of important data. That data is now lost to the corporation. As a result of the loss of that data, the corporation begins to lose money. The system has failed as a result of its refusal to diffuse responsibility and knowledge past this one man.

B. A Presidential Impeachment

The President of the United States is impeached for high crimes and misdemeanors against the people of the country. The system has failed, although it has survived through the fail-safe mechanism of the impeachment process. Why, then, do we say it failed? Because it failed to control its highest functionary, through apathy, overconfidence, panic, and so on.

Levels of Decision Making in Enhancement

The levels of decision making during enhancement are essentially the same as they were during the developmental life cycle, although the impulses originate in different places. Generally speaking, the request for enhancement is directed by the user, not to management as before, but to A&D, which has become a working liaison between the user and system development groups.

Frequently such a request originates during the postinstallation review, as we mentioned during the last chapter. Even when it doesn't, however, it frequently originates at somewhat lower levels of both organizations. It then becomes something of a struggle to organize a real systems effort, since the impetus to act comes from a different direction than usual. The organizations at the beginning of enhancement look something like the one in figure 10.3.

The system development group is then faced with reordering procedures to effect an organization that looks like figure 10.4. If such a reordering isn't accomplished, both user and system development groups run the risk of patchwork solutions and premature death (read: failure). Once the essential reordering is accomplished, the levels of decision revert to the order detailed in chapter 2, with the single proviso that A&D frequently has more discretionary decision-making power than it might have in a virgin effort.

FIGURE 10.3

Organization at Beginning of Enhancement

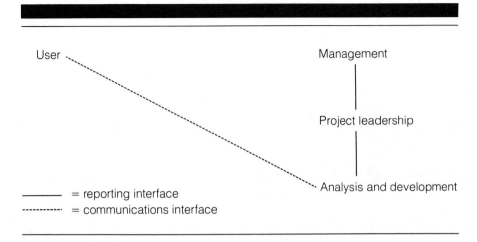

FIGURE 10.4

Organization at Beginning of Effective Enhancement

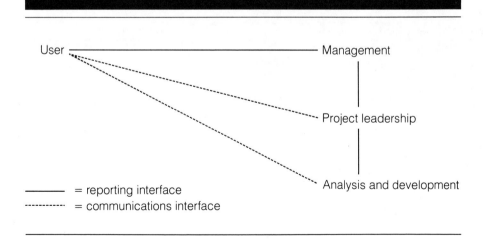

So that's it?

As far as any one system goes, that's it. We've gone through every phase and subphase, every procedure and document that we intend to.

We're not quite finished, though. Which is why you may have noticed that there's another chapter following this one. Take heart. It's short.

"Welcome to Ghirardelli Square," this mermaid seems to say. The old, outmoded chocolate factory has been enhanced to become one of San Francisco's prime tourist attractions.

Exercises

For each of problems 1 through 5, give examples of user, external political, competition, and legal enhancements to the system.

1. Heart pacemakers.
2. Airport operations in a large city.
3. Sanitation landfills (dumps).
4. Payroll systems.
5. Hospitals.

In each of problems 6 through 9, a system is given along with several enhancements. Classify each enhancement as to type. Identify which could be considered maintenance.

6. An accounts payable system. Getting an additional report based on volume of business. Reporting to the government on miscellaneous income.
7. An urban bus system. A federal law requiring reinforced bumpers. A state law on the emissions of nitrogen compounds from diesel engines. A request by drivers for air conditioning. A suggestion by the truck maintenance group to standardize on the type and number of buses in the fleet.
8. A government agency sponsoring basic research. The award by a well-known senator of a "Turkey of the Month Award" to one of the research contracts.
9. A request by a university for a new type of grant. The results of an audit by the Government Accounting Office that funds have been misspent.

For each of the systems in problems 10 through 14, decide what conditions would make you want to kill the existing system.

10. B-1 bomber.
11. Arms aid to developing nations.
12. Disarmament talks.
13. Busing of schoolchildren.
14. New brand of soap.

In each of the systems in problems 15 through 19, discuss how enhancements could make the system inflexible and unwieldy.

15. Automobile.
16. Plumbing system in a house or apartment.
17. Dam construction along a major river.
18. Survey project on cancer victims.
19. Funeral parlors.

20. Your government is currently supporting an unpopular regime in a developing nation. The ruler has asked for more sophisticated arms in return for his political support. What political environment internally would make you cut off aid? Increase aid?

21. Compose four distinct systems from the following list of changes. What do enhancements reveal about the original decision to build the system?
    a. A desire to change report headings.
    b. New laws governing reporting for unemployment.
    c. The demise of a part of the user organization.
    d. Your competition has cut its prices for an item 20 percent below yours.
    e. The price of fuel has just increased 20 percent.
    f. Demands for increased density of buildings.
    g. A change in the user organization requiring more analysis.

22. Select any system of your choice. You are in charge of reviewing the maintenance and enhancement work that has been done on the system. What criteria will you use? How can you tell whether the work has been done? What data will you collect?

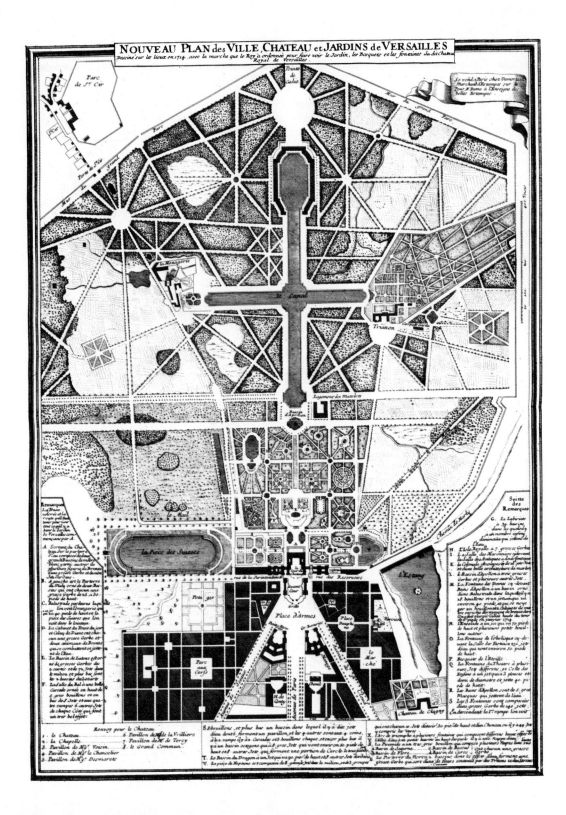

11

Systems in the Real World

We've lived through the system life cycle and we've seen what should and what shouldn't be done in using the systems approach. This last chapter concentrates on what is rather than what should be. As many of the readers of this text already know, the world is a place of contradictions. Things are very seldom what they seem to be—and frequently not what they should be.

The systems approach is both a theoretical approach and a practical approach to life. In that respect, it's very much like some other disciplines: engineering and social sciences. All these approaches attempt to build a framework by which to view life and to change it. We believe the systems approach offers singular, valuable benefits.

Cost

Almost everything has a price tag. Certainly everything in the world where people are employed has a cost. The social world can be seen usefully as a vast marketplace, where things, services, and ideas are bought and sold. And, while these costs may not be numbered in dollars in certain cases, they are always numbered in trade-offs of some kind. If you eat a hot fudge sundae, you eat it at the cost of 1,200 calories. If you put a window in the south wall, you put it in at the cost of the lost wall space.

Although no approach can offer a way to avoid these trade-offs, the systems approach does offer a way to become more conscious of them and to value them in a way that helps us to choose the best of two possible worlds. In the business world, for instance, the systems approach, when properly applied, can yield valuable insights. It can project and keep tabs on the monetary cost of new ideas, new products, and new ventures. It can help prevent overinvestment in projects that don't promise long-range profits. It can help increase efficiencies, help develop markets, and help keep track of data.

In the world of social science, where everything is intangible except the paper data and the people involved, the systems approach is especially valuable. Trade-offs in social science comprise perhaps the single largest headache, outside the subjects of study themselves. If you select one wording for a questionnaire, you select it at the expense of all other possible wordings. If you decide on a site for a recreation center, you forfeit all other sites. If you devote a grant to one line of investigation, you can't devote it to any other. If you study people in groups, you can't see them adequately as individuals. And so on.

Awareness of these costs is the first great advantage of the systems approach. By insisting at every stage of the game on constant evaluation and reevaluation, we hope to keep paths from becoming too restrictive to vision. By never committing resources beyond what we know can happen next, we hope to keep the paths from straying from the most productive avenue of exploration.

In other words, there's a cost for everything. You're better off to know what that cost is before you invest. You stand a good chance of knowing that cost if you apply the systems approach before you leap.

The Means to an End

In this world of limited resources and limited time, the second great advantage of the systems approach is that it keeps the goal in sight at every step of the way. From the time a problematic environment is first analyzed and a sustaining environment is first visualized, the systems approach allows no deviation from the search for and development of an applicable solution.

There are valuable by-products to every project, and the systems approach doesn't cut down on reaping that harvest of extras. However, it does keep the system development team from wandering into Dorothy's field of red poppies and falling asleep on the job. However pleasant chasing rainbows may be, it doesn't (in most cases) get much done toward specific goals.

With the sustaining factors and the user requirements firmly in hand, the problem solver (and his team) know where they're heading. And, if they can support the cost, they'll eventually get there.

Accountability

Most people can plan to be called by someone to account for what they do. The social scientist must answer for his funding and for his time. The engineer must account for his design and for his budget. The businessman has to account for—well, everything he does.

The need to account (unlike accounting, not just for money) points the way to the third advantage of the systems approach. The systems program leaves a clear, retraceable path of ideas, documents, and milestones. The system development team that operates properly never has to worry about explaining where it's been; it has only to pull out the files and look.

Analysis and Improvement

Nothing is 100 percent peaches and cream. No project, once completed, should have its file closed and be forgotten. Everything we do offers valuable lessons for the future. No approach yet developed to solving problems offers more potential for this process of self-evaluation than the systems approach. Because of the clear track it leaves and because of the logic that the system must have to progress, it provides ready-made tools for analysis and improvement at project's end.

Every system contains its own little (we hope they're little) failures. Those failures can be turned to future successes by a backward glance. The systems approach offers the invaluable resource of reliving our mistakes and learning from them. When you can grab that file and find out just where you made the investment that drove the final cost sky-high, you can avoid making similar investments in the future.

In a very important sense, this function of analysis and improvement is the most valuable role the systems approach can play in our lives. There are other ways of building a bridge, of taking a census, of planning a transportation network, and of buying a car. There's no other way that offers instant replay the way that systems do.

Reality Versus Theory

We said we were going to talk about what happens as opposed to what ought to happen. Here it is. The systems approach would be a worthless way of seeing things indeed if it were as rigid as many people think it is. It isn't rigid; it must be the most flexible way of thinking ever devised. In those cases where it becomes rigid or unseeing, it's no longer the systems approach. It's then a thinly disguised paper bureaucracy.

What do we mean by that? Well, all sorts of things, but let's take some examples:

Example 1. We know now that every system life cycle must pass through seven stages: initiation, feasibility, analysis, design, building, installation, and operation. Now that we know that, we can destroy that, because there are many systems that zip through stages so fast that we barely have time to think. When you decide to buy a new dishwasher, you've finished the initiation and feasibility stages by the time you've made that first decision. You actually begin work in the analysis stage—looking over the range of dishwashers available and investigating your finances to see what fits best.

When you throw away a ballpoint pen, you eliminate the maintenance and enhancement part of the operation because there's nothing left to maintain or enhance. You just buy a new pen. When you decide to install a new computer, you probably won't have much to do in the building stage because the manufacturer already did that.

The point is that while every system passes through most of the stages many systems need not take notice of the time they spent going through them. Certainly no one is interested in the mental feasibility study required to tell you whether you ought to have a green or a yellow vegetable with your lunch.

Example 2. We've made something of a point about documentation during the life cycle. We've specified several documents, including a set of user requirements, a project plan, a project summary, and a postimplementation review. Now we're going to tell you that you can do without them.

The real world isn't going to require them, at least not all the time. If you spent time documenting every system you live through or work through, you'd spend your whole life writing reports. Obviously, this documentation is only intended as a set of guidelines for important pauses in the life cycle. Each document represents a time of thought, a time of decision, or a time of analysis.

We call the systems approach just that: an approach. It's not a magical set of incantations that must be followed to the letter to successfully summon demons. It's a way of thought that is structured, deliberate, and thorough—a pattern of thinking that allows the problem solver to impose order on a disordered world.

What Can Systems Do for You?

A systematic way of thinking (that's what we're talking about) can give you a path to follow in solving your problems. Systems don't give advice, but they do give structure. They can't help you pay your bills, but they can help you decide whether to declare bankruptcy.

A systematic way of thinking will help you get to the core of problems more quickly than you might without it. When you keep your focus clear and your trade-offs conscious, you can make decisions more quickly and more intelligently.

The systems approach can help you pin down a hunch. Somewhere inside you, you just *know* that company A is a better long-range investment than company B. Why? Everyone else thinks the two are about the same. What do you do? You set up your own criteria, decide on your own yardsticks, and find out whether your hunch is just the product of last night's lasagna or an insight that may be valuable.

The systems approach can help you project a budget or a schedule because it takes everything in order and one step at a time. Once you find yourself at the end of a stage, you know where you want to get by the end of the next one.

The systems approach can keep you from chasing wild geese. Because you're constantly evaluating your position, you stand a better chance of quitting while you're ahead than the next guy does. You understand the

As any sand architect knows, there's only one spot on a beach that's suitable for such a project. Your analyst must find that one perfect place—where the sand is damp enough to build with and the tide is far enough away to leave the project intact.

difference between death and failure, and other people frequently don't. You're willing to cut bait when it becomes obvious that you're not going to catch any fish.

It would be impossible to enumerate the ways in which systems can be applied, just as it would be impossible to list all the problems in the world. And there are situations where the systems approach is useless.

Times to Forget Systems

Like anything else, there are times when systems are best left where they lie:

1. Forget the systems approach when you're trying to justify something that's already decided. You can't construct an information system that will convert a Democrat into a Republican, or vice versa. You can't construct a valid system to explain your income tax calculations to an auditor; you can construct something that *looks like* a system, but you may be better off without it.

2. Forget the systems approach when you're discussing matters of opinion or faith. There's no system that justifies religious beliefs or that leads to immortality. While that is probably overobvious, its corollary is that there's no system to help you stop smoking or to prove that Francis Bacon wrote Shakespeare's plays.

3. Forget the systems approach when there's simply no data available. The systems approach will be of no help in deciding where the abominable snowman lives or whether Atlantis is submerged off Bermuda or Brazil.

Just what is the systems approach?

The systems approach is a phased approach to the process of problem solving. It's a series of milestones and commitments. It's a way of thinking. Figure 11.1 enumerates the physical apparatus of the systems approach, detailing the stages, activities, end products, and possible sources of failure. This figure may be used as a map to guide you through system development. It may be used as an organizing factor for your system files or as an aid to devising a format for system development in your industry and/or organization.

FIGURE 11.1

Activities, End Products, and Possible Sources of Failure by Stage

STAGE	ACTIVITIES	END PRODUCTS	POSSIBLE SOURCES OF FAILURE
Initiation	Define problem and initial user requirements Determine initial group I and II lists Determine initial set of alternatives	No major document	Misunderstand problem
Feasibility	Collect data Analyze existing system Define alternatives Evaluate and select alternatives Update and review group I and II lists	Feasibility study	Collect wrong or incomplete data Perform erroneous analysis Select wrong alternative
Analysis	Develop project plan Determine detailed user requirements Determine system requirements Update and review group I and II lists	Project plan System specifications (user and system requirements)	Develop excessive or deficient requirements Fail to compare and relate adequately user and system requirements Underestimate costs and schedule
Design	Update project plan Develop logical design Develop physical design	Logical design Physical design	Develop incorrect design Fail to satisfy user and system requirements in design

STAGE	ACTIVITIES	END PRODUCTS	POSSIBLE SOURCES OF FAILURE
Design	Evaluate designs against system specifications Review and update group I and II lists		Fail to reflect logical design in the physical design
Building	Update project plan Develop and test pieces of the system Integrate and test the pieces Perform system tests Review and update group I and II lists	Completed system Procedures on using and operating the system	Perform incomplete tests Prepare procedures that are difficult or impossible to follow Fail to take corrective action
Installation	Update project plan Perform user training Perform acceptance testing (by user) Install the system Review group I and II lists	Accepted system Project summary	Inadequately train users Fail to cut over to new system
Maintenance and Enhancement	Maintain system according to system specifications Perform a postinstallation review Enhance the system to meet new requirements	Maintained and enhanced system	Introduce new errors in system Fail to perform needed changes to benefit the user

The systems approach is a refinement of the classical scientific method of observation and analysis which helps the scientific method apply to real-life, less-than-concrete situations. It's a view of the scientific method that helps us build, rather than just organize and analyze.

The systems approach is nothing more, nothing less, than a method of transforming problems into solutions. It all boils down to that. If you can keep your group I list in sight and successfully transform it into a group II list, you've got it made! The formula is simple:

> *problematic environment + system = sustaining environment.*

And it took us this whole book just to say that.

Photo Acknowledgements

166   Bulloz

170   Bulloz

171   James Motlow, Jeroboam

177   Bulloz

180   Bulloz

181   Marc Riboud, Magnum

186   William Rosenthal, Jeroboam

192   Largilliere painting from the Wallace Collection, London

196   Ciccione, Rapho

199   David Moore, Black Star

211   Bulloz

216   Photo Verney, courtesy La Documentation Francaise

221   Brown Brothers

228   Phelps, Rapho

232   Kent Reno, Jeroboam

240   Elizabeth Crews, Jeroboam

244   Le Coguen, Rapho

(Frontispiece page ii and chapter openers: Dr. Fisher Kimball, *Thomas Jefferson, Architect* from the New York Public Library)

(Endpaper illustration for the case edition courtesy The Newberry Library)

Index